BURIED ALIVE!

THE UK CABLE REVOLUTION

IAN SCALES

BOWERDEAN PUBLISHING

First published in 1996 by the Bowerdean Publishing Company Ltd,
8 Abbotstone Road, London SW15 1QR

British Library Cataloguing-in-Publication Data.
A catalogue record for this book is available from the British Library.

ISBN 0 906097 69 X

Typeset by Ingraphics Ltd, Wellingborough.

Printed in Malta by Interprint

To

Debbie, Allegra and Nyree

ACKNOWLEDGEMENTS

Many thanks to the all those within the cable industry whose views and knowledge have formed the background for this book. In particular: Richard Woollam and Niall Hickey, formerly of the Cable Communications Association; Mike Prymaka, Rowena Gardner, Stan Bihaun and Hugo Davenport at Cambridge Cable; Stephen Davidson and Ian Hood at TeleWest; Russell King at NYNEX; Alan Hindley, Nicholas Hyam, Henry Forde and Karl Edwards at CableTel.

Many thanks to Barry and Eddie Lloyd at Barrington Lloyd Associates and special thanks to my publisher, Robert Dudley, for his enthusiasm and encouragement.

Contents

Preface

Buried Alive!

Buried Alive! is about one of the UK's least understood Information Technology (IT) sectors – the Cable Communications industry. After long-awaited consolidation, the cable industry at the time of writing comprised 8 Multiple System Operators who in turn own or part own cable systems serving franchises awarded by the UK's Independent Television Commission (ITC). But multi-channel TV is now only part of the story.

Since 1991 cable operators have successfully added telephony to the service line-up and are now beginning to enter the Internet access market.

The UK is therefore a unique communications laboratory, and the on-going experiment is a valuable source of data for the rest of the EU and North America where local telecoms competition is also being introduced and cable operators are being viewed as potentially important players.

This book explores the current and potential power of the digital cable network in a world where the old 'analogue' systems which dictated sharp demarcation lines between industry sectors are breaking down. Digital systems – in publishing, TV production and distribution, voice communications, information storage and distribution, retailing and financial services – have blurred those comfortable boundaries. The UK cable industry, with its feet planted firmly in both telecoms and entertainment, is therefore a 'need to know' topic for an increasing number of people at all levels in a broad range of converging sectors – the computer industry, electronic equipment, Internet community, software, PC games, telecoms, data communications equipment, television, television production, advertising, print and publishing, local newspapers, and of course those working in the cable industry itself. We expect most readers will therefore approach this book with a mix of business need and personal interest. We have not attempted a technical tome or a definitive business report.

Buried Alive! explores the issues, explains how the technologies work and what the UK's cable network capabilities might mean in business and social terms.

1
Introduction

Nearly all the outstanding IT corporate success stories involve outrageous good fortune. Bill Gates, for instance, founder and CEO of Microsoft, was famously the man who was 'in' when IBM came by 16 years ago, shopping for a microcomputer operating system for its PC. Gates had the mother of all lucky breaks and has since been able to capitalise on it, but the Microsoft story is by no means unique. The whole, broad IT field in all its guises – equipment, software, services – is littered with 'brilliant' business moves involving more hindsight in the telling than foresight in the original investment – a fact evidenced by their often being unable, unlike Microsoft, to repeat the trick when it comes to the follow-up product.

How can it be otherwise? Technology products and services take years to develop from scratch, while market opportunities often materialise as a 'window', to use a for once appropriate piece of marketing jargon, briefly opening to admit the agile. Being in the right place at the right time – and often having a product or a capability originally designed for 'x', but now perfect for 'y' – is often the factor which sets those corporate success wheels in motion.

This serendipity factor has already smiled once on the UK cable industry. The late 1980s saw cable in the UK struggling – unable to convince more than 15% of 'passed' homes – homes whose street has been cabled – to sign-up for multi-channel television. At this time, too, it faced stiff competition from direct satellite – things looked bleak.

Then serendipity struck and almost immediately the industry's prospects transformed. Overtly encouraged by the government into the UK telecoms market in 1991 with the end of the country's failed experiment in telecoms 'duopoly', by late 1995 cable telecoms revenue had outstripped cable TV revenue. Cable's real assets, the network of ducts and their associated way leaves originally assembled to deliver the 'x' of a television cable, had been on-hand to provide the 'y' of a competitive telecoms 'access' network. The window was created by the regulatory realisation that what is called 'infrastructure competition' – where different networks compete to reach customers – was a pre-requisite for real telecoms competition and lowering prices.

With this in mind, it was obvious that cable could quickly reach the places other competitive telecoms operators couldn't, or wouldn't, reach.

With its new arm the cable business subsequently attracted massive inward investment from both foreign telecoms operators, anxious to learn about life in a competitive environment, and foreign cable operators, keen to see how telephony works in a cable business.

Another lucky break?

Now, in 1997, the cable industry may be on course for an even bigger prize – it is well positioned to enter another potentially lucrative 'window' as a major deliverer of interactive digital services. If and when it does so, it will have completed what may in future look like a brilliant outflanking manoeuvre – from a base in broadcast entertainment delivery it has gone on to build an important business in 'switched' communications – allowing customer 'a' to connect directly to customer 'b' to exchange information – in this case mostly to just talk over the telephone.

It is now poised to bring these two capabilities together to provide a new generation of communications services – involving what we currently understand as communications, entertainment, shopping and commerce – to attempt to become the 'preferred pipe' into the home and business as the industry jargon would have it.

This book sets out to explore two things. It charts the almost invisible growth of what has become by any standards a major UK business sector, employing directly and indirectly close to 30,000 people. In typical British fashion about the only time that cable enters the public consciousness is when it is a digger of roads, destroyer of trees, or when squabbling with BT over competitive practices.

In fact, the UK cable industry has helped to incubate a thriving UK telecoms industry which will go on to develop and sell a myriad of software, products and expertise to the rest of the world and, particularly, to emerging competitive telecoms markets in the rest of the EU as its members follow the UK lead in introducing telecoms infrastructure competition. In many cases this competition will be conducted through existing or fast developing cable networks.

Most importantly, *Buried Alive!* will explore the future possibilities.

This is really what cable investment is hedged against – a belief in the inevitable emergence of a third business arm involving a set of powerful interactive services. These new services will eventually subsume both broadcast entertainment and fixed telephony as we currently understand them today.

This general 'view' of the likely importance of interactive digital services, networked multimedia, the Superhighway, video-on-demand, and all the other 1990s buzz phrases used to conjure up the idea of computer technology put to work across a network, is also shared to a greater or lesser degree by television and satellite companies, the on-line information industry, and of course, the incumbent telcos such as BT. They all, for both positive and defensive reasons, have ambitions to carve out a chunk of the action, whenever and in whatever form it arrives.

There are, I think, three telling features of the cable industry which will see it compete effectively against its equally determined competitors.

The technology

Least importantly, it has the sort of network upon which first generation high speed interactive services can be built today. I say 'least importantly' because ultimately both BT's network and other, perhaps radio, networks will eventually be able to achieve similar capabilities – albeit through the use of different technologies. But the cable network looks able to deploy a relatively 'cheap' way of delivering high-speed data, and therefore to support high speed Internet access, through an intermediate technology inadequately described at present as the cable modem. This device will not be the ultimate answer, but it does put the cable network architecture on the starting grid for interactive service delivery while the competitive alternatives are either in the pit lane or still on the drawing board. Being first is always a major advantage.

The scale

More importantly, and perhaps more contentiously, the way the cable industry is organised may be an advantage. The last year or two have seen the emergence of a more refined understanding, by the IT industry as a whole, of the way 'interactivity' will develop. Not as a series of pre-designed, deliverable services controlled by a technology company, but through an organic process of trial and error driven by both large and small organisations and, of course, individuals. The laboratory is the network itself and the Internet will almost certainly remain the platform for this evolution.

For historical reasons, cable companies have been organised around local communities. Even as they consolidate into regional telephone companies, they remain relatively lean and decentralised – just the model of company that seems to be performing best in what business theorists are now characterising as the 'new paradigm' thrown up by the run-away success of the Internet's premier application, the Worldwide Web.

As I hope this book will show, what might appear to be a major disadvantage in terms of scale and scope when pitted, service by service, against one of the UK's largest companies, is all the time providing a less tangible but none-the-less telling advantage in terms of customer relationships. In the business market, cable companies have room to duck, dive, dodge and weave. Or as they would prefer to describe it, they are able to use their technology to 'meet customers' needs' on an individual basis.

As I shall also show, the underlying digital telecoms technology has grown increasingly flexible and, as a result, there is a huge growth in the range of services capable of being deployed. Where telecoms used to involve just leasing lines or point-to-point circuits, it can now involve all manner of complex arrangements – Centrex, ISDN, managed networks, call forwarding and voice mail – all deliverable from the operator's switch. Operating close to the customer and having a high degree of local autonomy gives cable operators the ability to make the most of the possibilities.

Where BT is able to engineer high-powered strategic alliances and has the resources to pour money into technical development, cable companies are able to build tailor-made alliances on a very local basis. These local relationships will increasingly involve putting local organisations and businesses 'on-line'.

This may be key.

The Internet currently conjures up images of a global wired society, but this aspect will be diluted as a more representative sample of the population begins to use it. As I look out my office window I see dozens of ordinary people engaged in very ordinary, everyday things – all rooted in local relationships involving the bank, the school, the library, the local supermarkets, local sporting and social events. If the Internet, or whatever it which support these activities. And it won't demand that customers must become computer nerds either – a major effort is currently under way to make consumer items connectable to the Internet, most obviously the TV set and the hand-held electronic organiser.

My neighbours, when they eventually do start using it, will change its nature far more than it will change theirs. Their Internet will give them access to their bank accounts, will enable them to order a taxi, book a ticket for the football, inspect local school prospectuses, deal with the local council, view films and watch local news.

Yes, they may also access entertainment services available on a global basis, and their children might seek information from all over the world for their school projects – but the global, cyberspace veneer currently associated with 'the Web' will exist only at the margin.

The package

Last but not least, cable is in pole position because it already delivers 'content' in the form of increasingly specialised TV channels. One of the most striking trends at present is just how quickly important aspects of both the broadcast TV and Worldwide Web businesses are converging. The marketing battle between Netscape and Microsoft for the Web browser market is about the content on offer behind the branded Web browsers as much as it is about the software itself. Meanwhile, content specialisation must become more important in broadcast where a single channel will soon be fighting up to 300 other channels for eyeballs.

On both sides, the technology is forcing the provider to compete by defining the target market more tightly. As I will show, some specialised content providers are already beginning to view satellite/cable TV and the Worldwide Web as increasingly similar ways of delivering the same content to different customers. This convergence will enter a new phase when the Internet is able to deliver video in real time, just like a cable channel. At present video clips are captured from the Internet (downloaded) and played later as it may take say 10 minutes to download one minutes worth of moving picture.

Developing Internet access alongside a growing package of channels will put cable operators in an exciting position.

Dreaming up the future

According to historian, Joseph Schumpeter, "The success of everything depends upon intuition, the capacity of seeing things in a way which afterwards proves to be true."

Schumpeter was not advocating alternatives to empirical understanding – rather the opposite. His definition of intuition is really super-empiricism – the ability to make sense of masses of disparate data in such a way that it points to an indistinct but none-the-less solid shape in the fog. As Schumpeter suggests, intuition must take over when the variables are far too numerous and their specific effects too uncertain to attempt a water-tight 'proof'. It is an intuitive ability the Gateses of the world share.

As the cable franchise ownership is consolidated into two or three major groups, the industry will have the scale to negotiate effective interconnect and distribution agreements with the powerful media and communications giants who share the same pond. But the local focus of the cable industry will remain. Its ability to organise itself in such a way that it maintains its community focus will still be key to its success as an alternative business model in the communications field. BT will always be a national and international entity – the cable industry now has the ability to strike a new and effective balance between global power and customer intimacy.

In the struggle for understanding now facing those in the information or entertainment business – as they view next-generation media and ponder their companies' best strategies – it boils down to understanding the power of the 'thing' in front of them. Not just how it operates and how powerful it is as a 'system' or as a technology, but in terms of the way it is likely to react to and in turn influence the chaotic forces loose in the wider world.

The rationale for investing billions of pounds in building an advanced network past much of the UK population is a bet on the ultimate value of that shape in the fog.

2

The Vision Thing

Whatever happened to Interactive TV and 'multimedia convergence'? In late 1993 much excitement was generated by the press and television by what we were told was imminent arrival of a 'new media'.

The details were unclear, but to the 'lay person' reading one of the Sunday newspapers, it all appeared to hinge on new technologies. One day soon, we were told, a corporate giant of some description – it could be a telecoms operator, a television company, a cable company or a new conglomeration of all three, with maybe publishing and on-line information companies thrown in for good measure – would be able to package a range of new technologies involving high speed processors, broadband digital services and video compression. This technology would be delivered as a new, 'interactive' generation of services and would change all our working and resting lives.

But it wouldn't stop with a 'new media'. All the old media – TV, films, newspapers, periodicals – were due to coalesce in a new 'converged' offering which could deliver network-based multimedia where text, moving and still images could be mixed and, most importantly, interacted with. In one form or another, users would become active selectors of specific material rather than passive viewers.

Some elements of this vision are beginning to take shape, but there has been one big important difference – those multinationals who thought they were going to be in pole position on the superhighway have not, at least at this stage, emerged as the key players in the process.

Neither have the huge corporations controlling global publishing, television, film making and its distribution, radically rearranged themselves into new power blocks – or at least, not yet. Somewhere along the way, someone or something seems to have changed the script.

Back to the drawing board

While the global communications giants were imagining the power and profitability of interactive TV, a much larger, and much more successful, service trial was developing on the Internet.

Over the past three years the Internet's Worldwide Web (often just called the Web) application has succeeded in taking the spotlight on interactive services, and so far the running here is being made, not by a corporate giant which looks suspiciously like a large telephone company or television corporation, but by a plethora of relatively small companies – each dragging their own specialised piece of a very large jigsaw into position. This 'open' network model seems to work in a way that the telcos' finished service delivery model has not.

For while the Web was signing up millions of participants (both content providers and consumers) worldwide every month, the initial burst of interactive TV trials initiated about three years ago by both telcos and large US cable companies, ran out of steam.

In all cases, although the delivery technology worked pretty much as expected, the users of these services failed to respond in a wholehearted way to the offerings – usually movies on demand, home banking and some home shopping. What was missing was the subtle chemistry present in any successful medium – where specific user demand and innovative content meet to generate a virtuous circle.

In the event it now appears obvious that these trials were testing the wrong thing by concentrating on consumer behaviour on the one hand and whether the delivery technology would work on the other.

The success of the Web has showed that the boundaries of a real trial are much wider.

As a result, it now looks as if new interactive services will evolve, not as a result of technologies being put into place, but as a 'slow build'. Information providers at one end of the networks which deliver the information, and users at the other, will progressively work out for themselves what sort of services will appeal, how they should be presented and how much should be paid for them. Those all-important network technologies which excited all the attention in the first place are now being viewed as simply the bit in the middle – the stepping stones for the development of interactive services, not the enablers.

The rise of the Web

By mid-1994 the Web, and the number of people signing up to access it, was growing exponentially. Dismissed at first by the telecoms establishment as a blip on the radar screen – the networking equivalent of ham radio – it nevertheless continued to develop, and the curve still shows no sign of flattening off.

The phenomenon of the Web has inspired all manner of bandwagon jumping. Doom-laden scenarios (pornography rampant, the exchange of bomb-making information etc.) are being predicted at one extreme, while utopian visions of a wired community where everyone learns tolerance and national boundaries are broken down, are being peddled at the other.

But the real lesson of the Web is not that one technology is better than another, or that new communities fostering good will and enlightenment will arise in its wake. The real lesson is that the successful development of interactive services is all about putting content in touch with users and making the bit in the middle, the network which moves the data, as simple and as invisible as possible.

This is an inversion of the traditional approach to on-line services which, until now, have designed from the top down with the engineering stress falling on security and reliability. Despite bullish forecasts, this on-line service model has never experienced the growth forecast for it.

Now even the large telcos like BT are slowly waking up to the fact that the world has changed and that either the Internet itself, or at least the business model that the Internet has unleashed, will define interactive and on-line services.

What's it all about?

So what are the key elements of what management-speak likes to call a 'new paradigm' which the Internet seems to have revealed?

The Internet is simply a network of networks linked by a relatively simple set of protocols and conventions. The key to the whole set-up is an agreed global addressing scheme which allows one computer to find another on the network. With this in place, the underlying Internet Protocol (IP) sets some basic rules for how data between computers can be exchanged. And really that's about all there is to it.

With this equivalent of an on-line operating system in place, applications are then built on top. The Worldwide Web has proved to be the most powerful of these applications due to its ease of use, and the ease with which information can be provided across it.

Once connected to the network the Worldwide Web user navigates between documents which can be stored anywhere on the Internet. The documents are simple text documents with embedded commands which are understood by the user's browser software. The browser requests the file and displays the text and its embedded graphics using its local system's fonts and facilities.

The browser makes full use of an MS Windows/Macintosh style user interface. Users 'point and click' highlighted text or graphics to move to another document which may be the next in a series, or could exist on a completely different server in another country.

Even in the few short years of the Web's existence, all manner of enhancements have been developed to make it more exciting and useful. Forms can be accessed, filled in and returned; animated graphics can appear on a page; and the next step is to allow users to execute high-level programs and procedures from within a document using the Java programming language.

Yet more goodies are on the way as the Internet protocols are upgraded to support 'real-time' data traffic – Internet telephony and the delivery of high quality video will round off its capabilities for the foreseeable future.

Loosening it up

Suddenly the Internet seems to be in possession of the world's most comprehensively successful interactive services trial, and the trial data, all very obviously and publicly available, seem to hold two major lessons for the rest of the industry.

First the 'intelligence' which runs the interactive services seems to be happiest and most prosperous when it migrates out of the core network and starts running things from the edge. This, after all, is what really distinguishes the Internet from all the other data network schemes before it.

The Internet provides a flexible 'minimum' of facilities and constraints – and this allows this complex of competing technology companies, content providers, service providers and so on, to innovate and build new applications and new businesses. There are no 20 year cycles of standards formation, no EU- primed research projects and pilots.

This is not just a technical distinction, because the way the technology is arranged is mirrored by the businesses running underneath. The people building these new applications for the most part work in loose confederations of quite small companies. They compete and they co-operate; they innovate and some of them succeed.

One of the important sources of market power on the Internet flows from ownership of a 'platform'. A platform is intellectual property, usually software. Commercial power on the Internet comes from getting users or other Internet content or service providers, to use or incorporate your intellectual property – the Internet itself is the medium for marketing and distributing it. So in this environment, the conventional tangible business measurements – physical assets, distribution networks, current market share and so on, suddenly have little or no value.

Microsoft seems to have learned this lesson. It originally tried to build an alternative Microsoft network to compete with the Web, but has subsequently realised that power really flows from the browser software, not the network. It is currently engaged in stiff competition with leading browser provider, Netscape, to dominate the market.

Netscape itself became a multi-billion dollar corporation because the promiscuous dynamics of the Internet allowed it to boost the value of its intellectual property by the simple expedient of populating the entire net with its browsers and server software. It did all its marketing and achieved market dominance before even beginning to look like making a profit.

But it didn't just distribute the browser: it worked on building a coalition of mutual dependence with other software companies. Which brings us to the second lesson.

Starting cheap

Another strength of the Web is its ability to admit content providers without asking for huge capital expenditure. This has lead to organisations involved in all manner of information provision developing a Web presence just to learn how the medium might be used.

From this position, the idea of going on-line with commercial products no longer looks like being a major and expensive leap into the dark at some future point – more like a gradual evolution as the Web develops and the commercial possibilities become better defined.

One of the most enthusiastic Web adopters has been broadcast TV – Web addresses appear with TV programmes and increasingly on television advertisements. Hollywood has also been quick off the mark. Web marketing is already seen by the studios as a crucial pre-release activity for many categories of movie.

Technologies as stepping-stones

It has turned out that you don't create new interactive services by developing new technologies and spending billions of pounds deploying them. Instead you end up with new technologies as a sort of by-product from developing the new services themselves. The technologies are stepping stones along a path of least resistance towards individual consumer or deliverer preference. So the solution is not to provide a service, but to enable a development process. The services themselves will evolve organically.

The digital technologies that make it possible to dream up concepts like interactive television are not susceptible to corporate hi-jacking. Interactive TV is not a boxed-up deliverable like television itself – it can run on a PC or a network computer, or on a computerised TV. It can be delivered across the Internet or even, perhaps from a satellite.

It does not even require much agreement about what it is. It may be 'near video-on-demand' for instance (where several channels are used to stagger the showing of a single film) or it may be a broadcast sequence with an interactive component superimposed – like a sort of two-way teletext.

Arguments about 'service definitions' and so on are futile and miss the point. Interactive TV, to take just one example, can be whatever both user and owner of the content agree works for them.

This outrageous promiscuity affects every budding 'player' in the chain. Unlike the situation with previous media, there is no position to be staked out in the physical chain (owning an expensive printing press, having a license to operate TV in a portion of radio spectrum). Or rather, there is no position to be staked out which can't be bypassed or whose value can be determined in advance with any degree of certainty. There are just too many technology stepping stones available – if participants view one route to be slightly too expensive or too limited, a slight skip to the left or right will probably find another.

If this dynamic couldn't be deduced by analysing the available evidence, it can be seen in action on the Internet.

Instead of a value chain, the traditional way of getting a handle on the way a business sector works, the way electronic services seem to be working across the Internet is beginning to look more like a value matrix – an altogether more volatile and fascinating business model and one we'll inspect more closely.

Migrating the content provider

The UK cable industry, with its feet planted firmly on both sides of the converging broadcast entertainment/telecoms divide is therefore in an excellent position to take best advantage of the way the entire area seems to be moving. With broadcast TV channels proliferating and becoming more specialised, cable provides an extremely powerful final delivery vehicle. Not only does it – as a system – match satellite in its ability to deliver hundreds of channels, it provides the best way of migrating those 'passive' broadcasts towards a more active and interactive model.

Unlike conventional satellite, its economics will allow it to deliver very local content. More importantly, the network is potentially fully two-way. This could mean an increase in the level of interactive control viewers are able to exercise over their viewing. But it also increases the active control able to be exercised by the content or service provider in terms of the way a medium can be targeted and marketed.

The nature of this dynamic will be explored more fully in Chapter 11.

As networking companies with advanced, high-speed networks already in place, cable operators have the wherewithal to deliver high speed web access, electronic commerce, mobile/fixed telephony service integration, and interactive entertainment, as well as more and more channels of broadcast TV.

The cable business meshes with the new model for interactive service delivery. Not a 'big bang' where an engineer arrives from the telecoms company and installs a new interactive service, but a slow burn where competition forces content owners to migrate to new ways of distributing their information or entertainment to more and more tightly defined segments of customers.

3

Catching the Tube

The business of delivering multi-channel broadcast television was the original basis for the cable industry. As we have already seen, TV distribution has already been nudged into second place in revenue terms by telecoms, and it will almost certainly become an even smaller slice of the pie as interactive services become a serious business strand. To the extent that new services will compete directly with cable TV for residential discretionary spending power (in a way that telephony doesn't), the importance of the TV offering may well diminish even further.

This chapter looks at the cable TV arm of the UK cable business and outlines some of the threats and opportunities ahead.

First steps

Strictly defined, cable TV really began in the UK with the development of so-called 'rediffusion' services'. First introduced in the earliest days of television broadcasting, rediffusion overcame specific reception difficulties for clusters of subscribers back in the days when TV ownership came with many of the trappings of a full-scale technical hobby – much hands-on effort was often required by the TV owner to produce a reasonable picture, which is about where we are today with the Worldwide Web.

But UK Cable TV – in its multi-channel sense – was a child of the first Thatcher government. The then Tory technology minister, Kenneth Baker, hatched plans for 'wiring' the UK by offering cable TV franchises to companies willing to undertake the capital-intensive work required to get the systems off the ground.

Licenses were initially granted by the Cable Authority, established under the 1984 Cable and Broadcasting Act. With its replacement with the 1990 Broadcasting Act, the Independent Television Commission (ITC) took over the licensing arrangements.

The cable TV licences gave their holders the authority to seek 'way leaves' across public and private property and to offer broadcast cable TV services. The political objective was to build a cable TV network rather than to garner licence funds for the Treasury, so bidding was scaled against the applicants' 'build' plans (the rate at which they proposed to build their cable networks) and their general organisational worthiness.

The regulatory regime was light in comparison to the US model where the prices cable companies were allowed to charge for connection were set by the Federal Communications Commission. But in the UK, companies who won licences were expected to meet a series of build 'milestones'. Those that didn't deliver on their promised 'build' would be in danger of breaching the licence and losing it.

In 1990, however, just before the cable companies were allowed into the telecoms market, the UK government decided that the previous arrangement was on the generous side, given the commercial opportunities now available (and, presumably, with one eye on the political sensitivities of granting 'free' access to foreign companies) and introduced a conventional bidding procedure for the outstanding franchises.

More light than heat

In the early 1980s it was possible for Baker to tie the whole initiative in the public mind to the newly applied principles of fibre optic communications – the cable network would introduce a new high-speed infrastructure. It was the 'white light', rather than the 'white heat', of the technological future.

Even at that distant beginning, TV delivery was envisaged as a starting point – fibre would be good for us in other, then still vague ways to do with a technological future and high speed networks.

As things have now turned out, Baker's plans look prescient, but for much of the 1980s the industry was beginning to look like a bit of a black hole.

The rate of investment, and therefore build, was sometimes low. Even where there was build, the multi-channel TV concept seemed to get off to a poor start with operators having great difficulty building their all-important 'penetration' rates (the percentage of customers signing up once homes had been 'passed' by the cable network).

The extra channels available through cable were perceived by a large proportion of the target population to be irrevocably 'down-market'.

By January 1990, despite passing over half a million homes, the industry had managed to connect just 87,000 subscribers: a penetration rate of 15.6%.

Competition from satellite was both a problem and a blessing. While the development of Direct Broadcast Satellite (DBS) services had helped foster a feasible channel offering for cable to deliver, they (soon to be 'it' once Rupert Murdoch's Sky merged with the alternative service, British Satellite Broadcasting) were also proving to be stiff competition in the residential market, further stalling penetration.

The following tables illustrate, the penetration rate levelled off in the late 1980s to pick up only after the cable industry was licensed to offer telecoms services in 1991.

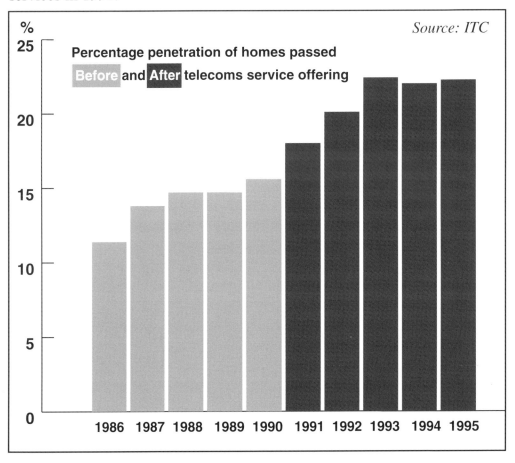

So far, it is fair to say that the performance of the cable TV side of the business has been disappointing. Cable penetration of just over 20% of homes passed (counting telecoms) seems to be the norm and all efforts to raise it substantially have so far failed.

However, with the introduction of telecoms service, the level of investment accelerated and this was reflected in the build rate (above) which also accelerated from 1991.

Options

Opinions differ about the best way to go forward.

As one step towards the long-term objective of raising penetration, cable operators are currently attempting to wrest some market power back from the programme providers who, from the cable operator's point of view, simply see cable as another form of broadcast distribution and expect them to act as a conduit for whatever is offered. Needless to say, the cable operators have a slightly different conception of how the dynamic should work. They want to build an attractive 'package' of channels to meet their objectives of gaining penetration to their homes passed.

Inside, the industry seems to be polarised around two points of view as to the best strategy to adopt on multi-channel TV and its marketing. The objective is somehow simultaneously to meet the penetration objectives, maximise revenue, and build a broad offering.

On the one hand there is the multi-channel universe strategy. This has as its starting point that the whole dynamic behind multi-channel TV is to give access to a universe of channels for a set sum. So much of the structure of the rest of the industry is built around this approach that it is hard to see how it can be modified.

On the other, some strategists champion the idea of presenting flexible packages with as little as perhaps five or six channels which could be branded to appeal to different interest groups. These 'bouquets' could then be packaged in innovative ways with both telephony and, perhaps, interactive services. It is easy, for instance, to envisage a package featuring Sci-Fi and music channels together with an interactive games service.

Fundamentals

On the other hand, programme providers charge cable operators to distribute their programming. This has been viewed as a happy and necessary arrangement in the past, since without the content the cable equation obviously didn't make any sense. But each extra channel costs the cable operator crucial extra pennies per home and this must be passed on to the subscriber.

For the cable operators the overall objective is to provide an attractive TV package with at least some degree of programming diversity. This way they will be able to increase that all-important penetration rate beyond multi-

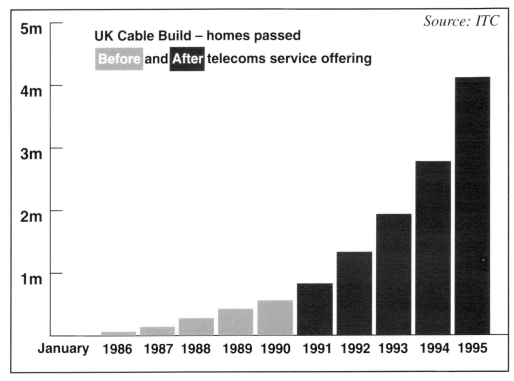

channel TV's core demographic constituency.

The programme providers, on the other hand, have a tendency to produce programming for the core, mass market. This is where they calculate the best audience shares can be won and where the best advertisers and advertising rates are available.

As the volume of programming grows the cable operators can afford to become more demanding as several elements in the market begin to swing in their favour.

First, there is a limit to the extent to which the programming costs can be loaded onto the subscribers' monthly bills, so negotiating the rates paid back to programme providers per subscriber passed is an increasingly protracted and cut-throat business. Cable negotiating power here has now been increased by the fact that several channels target the same viewer interest – for instance in music, sport, weather and so on.

The cable's capacity underwrites this logical limit, for while the subscriber may be reluctant to see the rental rise, the networks themselves are beginning to grow short of space. While cable networks theoretically have vast capacities, the way the technology is currently engineered means that most systems have an upper limit of around 50 channels.

So the days when programme providers could expect to be almost automatically allocated space have long gone. While programmers are anxious to build an identity and a niche for their particular channel or channel package, the cable operators must be concerned about multi-channel TV as a total package.

Part of this concern relates back to their objective to shift customer perceptions about what is on offer – in publishing terms, they would like to be able to move the image of the package 'up-market'. A good place to start in this effort is to ensure a 'diversity' of programming.

Muscle

There are now signs that operators will pursue this objective more aggressively than they have in the past. Telewest, for instance, recently decided to flex a little muscle over the introduction of (and the terms on offer for) a new package of programming made available from a joint venture between Granada Television, the leading Independent regional TV company, and BSkyB.

GSkyB, as the joint venture has been dubbed, offers a seven channel package (not all available 24 hours), and Telewest had calculated that delivering it under the terms on offer would load an extra £1 per month onto cable bills. Telewest decided to put the issue to its existing subscribers and ask them for feedback (using its TV channels to elicit a response). It told GSkyB that the customers would decide. In the event the issue was settled speedily – Telewest said that early indications showed that the customers were keen to have the new channels, but no doubt the price was also adjusted in Telewest's favour. Whatever the case, the move was a low-calibre warning shot fired across the bows of the entire programming sector.

Problems for satellite too

The under-demand for multi-channel TV seems to be endemic. Originally, Sky had forecast a 2 million early adopter phase which would come quite quickly. It did. But the growth peaked. By the mid 1990s there was supposed to be 6 to 7 million satellite TV dish owners. In fact by mid 1996 there were around 3 million and some experts suspect that growth has levelled off.

Although cable has passed only about 4.5 million homes (nearly a quarter of total UK households), it has achieved a much better penetration rate of just over 20%. As satellite passes all 21 million UK households its equivalent penetration is around 14%.

Even so, there is still a long way to go before the industry can achieve the 60% penetration level currently common in North America. At present it is possible to argue that the legacy of UK recession of the early 1990s has yet to be shaken off and cable TV subscription, definitely a discretionary spend for most households, has suffered as a result.

But even taking this into account, there seems little sign that penetration levels will rise significantly.

Where to?

Some hard decisions seem to face the cable industry over the future strategy for cable TV. Does it continue to push a single, take it or leave it price point? Or does it make sense to risk diluting the value of the business by looking carefully at packaging 'bouquets' of channels, or even single channels, at a range of price points in the hope of raising penetration overall?

Part of the problem may be the psychological domination of BSkyB in the market. BSkyB uses its position to force cable operators to take 'bundles' of programmes – a practice many in the cable industry view as anti-competitive.

The recent decision of the UK's Office of Fair Trading to recommend only minor changes in the way Sky conducts its business with the cable companies came as a heavy blow. But it may force the operators to reassess their long term view of the market. If it is increasingly difficult to mesh cable marketing strategies with the way Sky wants to package and price its offerings, it could force the industry to look again at where it wants to be in the long run. Does replicating the dynamics of satellite TV delivery across a potentially interactive network really make all that much sense?

Wrong model?

But apart from the business issues, many are asking whether it makes sense to continue trying to force a homogeneous, multi-channel offering onto the UK market at all.

There is an opinion within the cable industry that the UK publishing business, rather than the US multi-channel TV business, is a better source of clues on what may work best in the UK market.

The first immediately striking feature of a UK newsagent is the sheer range of product available – not just in large chain stores, but in any small town outlet. Every interest is catered for – from caravanning to pigeon fancying at one end of the scale (so to speak) to horse riding and magazines which explore country houses at the other. In between, there is a huge of range topics with more neutral social overtones, but which nevertheless manage to differentiate themselves. Such dynamics are at play in every market, but in the UK the degree of segmentation seems to be greater

And these things matter. A *Guardian* newspaper reader would rather suffer an amputation than be seen with *The Daily Telegraph* under his arm. A *Vogue* reader would never be seen on the beach with a copy of *Chat* .. and on it goes.

Terrestrial digital

The cable industry also faces new competition. Digital terrestrial TV will allow many more channels to be delivered to standard aerials and new spectrum has been assigned to support it. But all the technical steps forward that allow terrestrial digital TV, will also pay off for the cable industry. If more channels are made available through compression, then an equivalent multiplication applies to those same channels running across the cable.

Except that it is even better than that. Terrestrial digital and cable digital will both require a digital decoder box (hopefully compatible) at the subscriber's premises, so the capital cost of terrestrial connection starts to look similar to the cost of cable connection. The difference will be that cable service will be able to offer hundreds, not dozens, of channels.

Digital satellite

Digital television satellites are also about to provide digital television broadcasts, but being a digital medium they could also broadcast text, voice, still pictures and anything in digital format.

For the customer, digital television has a different screen aspect (the screen is wider) and the picture has greater clarity (although not to a startling degree).

Given the extra capacity there will be more scope for specialist 'broadcast' programming – certainly enough to dedicate to educational channels for schools.

All these advantages, of course, will also accrue to cable in a completely digital environment.

But despite the hype, digital satellite television does not provide a sensible basis for a fully interactive network – one where the viewer, sitting at home in front of a screen, could order up the transmission of some obscure French film, or pull some information off the Internet.

Such an interactive application involves dedicating end-to-end bandwidth between each user and a server. Three hundred channels may be a revolutionary broadcast opportunity – it does not form the basis of an interactive broadband service.

The confusion has arisen around what constitutes 'interactive'. The economics of satellite technology work only for broadcast applications – ie where many viewers are somehow simultaneously viewing information from the same channel, whether they realise it or not.

On this basis, digital satellite TV will be able to use some of its many channels to provide what is currently called 'near video-on-demand' for a limited number of films. Video-on-demand (VoD) is where the viewer can 'call up' a film or programme.

Near video-on-demand

Under this definition the delivery technology is sophisticated and powerful enough to provide what is essentially a private viewing for a given piece of content, by serving up a digital stream from a powerful 'video server' and sending that stream all the way to the viewer's TV set.

Near video-on-demand aims to provide the next best thing by staggering the start times of a single film across, say, ten channels. Viewers can exercise more choice over their own viewing schedule than is available with multi-channel broadcasting, and they can replicate fast-forward and re-wind by moving between channels.

If you envisage that most viewers will want to watch the same few films at any given time, near VoD makes a lot of sense as an interim service. As viewing habits further fragment, however, most observers believe that proper VoD will be viable.

But digital broadcasting from satellite could also be used to provide a high capacity teletext-style service, where the digital set-top box picks out the pages of text the viewer wants to see from a continually repeating stream of data. Very large amounts of information (in comparison to analogue teletext) could be cycled, within an acceptable average wait time, to provide news, stock prices, weather forecasts and the like.

Real interactivity using satellite – where users choose exactly what they want to view – would require a terrestrial return path of some kind (over the telephone or via one of the cellular networks, for instance). This technique could be used to allow viewers to call up data interactively, but the applications would have to be limited to a very light data requirement.

Satellite service may have a lower per customer capital investment cost, but it is unlikely to be able to provide full interactivity because vast though the satellite data capacities may seem, they are not nearly enough to give every subscriber their own high speed channel.

Again, in terms of sheer channel capacity, any compression technologies which increase the capacity of satellite delivery must, by definition, confer the same advantages on cable service.

Radio distribution

Another consequence of digital TV will be to make the concept of cable by radio more attractive. The idea of delivering broadband services by radio is not new – but the economic case seems to get better as more chunks of spectrum are brought into play. Enthusiasm is currently being generated by Microwave Video Distribution Systems (MVDS) designed to operate across 2 GHz in the 40GHz band – a segment recently released by the military.

On paper the approach looks persuasive.

It is claimed that the low-powered line-of-sight systems using very small antennae (smaller than a satellite dish at both ends) could deliver over 200 digital TV channels plus telephony and a return path. The signals would be identical to, and could therefore be descrambled by, BSkyB set top boxes.

It would operate as a sort of horizontal cellular satellite service. The low power, directional nature of the technology allows the frequency to be reused in cells (in a similar way to cellular telephony) freeing up some of the capacity for point-to-point, rather than broadcast applications.

These could include asymmetrical (where most of the data flows towards the viewer) services like high speed Internet access or VoD.

Frequency will also be available for telephony, including basic rate ISDN (see Chapter 5, Pandora's boxes); and capacity could be used to deliver DASS circuits or higher for business applications.

Similar technology is already being used in Hong Kong, at 12GHz, originally as a stop-gap for cable but then, as it seemed to work perfectly well, as a replacement. The US Federal Communications Commission (FCC) has allocated radio spectrum for its equivalent, LMDS – Local Multi-point Distribution Service.

The companies involved claim that the approach can price itself in at a substantial discount where cable fixed costs are high per household (where dwellings are spaced, or where there are small clusters of dwellings). They also point out that the approach replaces fixed costs with variable costs, since the majority of capital outlay only occurs when a subscriber signs for service, not when the initial build is put down.

This movement of costs from 'up front' to 'per subscriber' sees the cable radio business model work best when the current penetration rate (around 20%) is assumed.

Downsides

If the UK take-up were eventually to reach US levels of 60% or more (the long-term plan, after all) then radio looks less attractive as the long-term option.

In addition, the radio equipment depreciates faster than ducts and fibre and the delivery technology (which must be technically quite specific) might constrain the operator's ability to deliver some services – there would be no equivalent to dark fibre, for instance, where the cable company leases physical fibres to customers who then pump their own signals down them.

The obvious application in the cable context may be to service sparsely populated areas whose build economics have put them right down (or off) the priority list.

Work is progressing on a technical definition for the return path for MVDS and telephony standards are also expected.

4

Hello BT!

The development of Cable Communications networks in the UK is a bold experiment. It has put the UK in the vanguard of European communications deregulation and competition by commencing the world's first duplicated wireline local access network – the bit of the network which runs from the exchange to the subscriber's home or business.

For political, security and organisational reasons the tendency to place the national telephone system under state control has been almost universal. In most European countries local telephone companies, where they originally existed, were absorbed into the state sector as an arm of the postal service. Until recently, in fact, all national telephone organisations were known as a PTTs (Posts, Telegraph and Telephone). The development of the telecoms arm of the cable industry sees a return in the UK to the original model of the 'local' telephone company.

Small more likely to be beautiful?

Where the local telephone company model has been maintained it seems to have done rather well. The best European example is in Finland which has traditionally structured itself around a plethora of local companies. Perhaps it is no coincidence that Finland is widely recognised as the most 'telephonic' society in the world, with both fixed and mobile telephony usage at very high levels. As a result, its indigenous telecoms equipment industry, most notably telecoms equipment vendor Nokia, sees Finland punching well above its weight on the global stage.

In the UK telecoms users are beginning to get the benefit of proper commercial competition between telecoms and entertainment service providers. But increasingly, as regulatory and technical restraints are removed, what are now cable and telecoms networks will look pretty similar – both will be able to deliver voice, data and TV (interactive and broadcast) and the differences in the technologies they deploy to do this will be irrelevant from the users' point of view.

Leading by example

The UK cable industry's successful foray into telecoms has done much to swing policy opinion in the rest of Europe around to the idea that this 'infrastructure competition' in network access is both feasible and desirable.

Infrastructure competition is now accepted as a critical element in the creation of a properly functioning market for telecoms services – and the UK cable industry has done much to validate the theory in practice.

Even five years ago, the desirability of infrastructure competition was a contentious issue. Telecoms, it was argued, was a natural monopoly. How much more sensible it appeared then than now, that a single local infrastructure be maintained across which 'service providers' might compete with differentiated offerings – no duplication of effort, no multiple digging-up of streets.

But in practice, competition to provide bandwidth at ever-cheaper prices seems to be a necessary precondition to a virtuous cycle where falling prices stimulate new applications or a greater level of usage, which in turn stimulates more revenue, more investment and a further lowering of prices – classic economics, in fact.

But pressure for change through the introduction of market forces was always intercepted at the public interest level – which is why progress towards a competitive telecoms market in Europe has proved relatively slow. Policy makers feared that competition unleashed in the profitable sectors of the telecoms market – business and long distance – would slash the dominant operators' (BT, France Télécom *et al*) profitability and make them unable to provide universal service at affordable prices.

The UK experience has helped to illustrate that no such downward spiral need take place.

Far from falling apart under an onslaught from cut-price competitors (always the doomsday scenario) and therefore being unable to discharge its universal service obligations, BT seems to have thrived in the bracing new conditions.

Not only has it so far held up its profits by becoming more efficient and more market-driven, but it has demonstrably improved its level of service at the same time. Much of its strategic attention is now focused on becoming a major global player. It has a engineered a range of joint ventures, principally in the US, Germany and Spain, and is determined to build new internationally-focused businesses.

As it turned out, providing universal service was only a problem under the monopoly because BT's costs were so high. With reorganisation and the deployment of new technology BT has found it possible to reduce the costs of running the local network to the point where the vast majority of residential subscribers actually represent a profit. Those who don't spend enough to cover the costs of their connection to the network have been estimated to represent a loss of less than 1% of BT's profits.

The old routine

In Europe, however, the market forces argument is still far from being won at a national level, so it is worth looking closely at how the arguments are used to defend the status quo.

The way most national telcos had structured their businesses usually meant that revenue from long distance national and international calls, plus extra revenues from business users, effectively subsidised domestic subscribers.

According to the operators' own figures, without this subsidy universal service in developed countries could not be ensured. The alternative was to have cost-based service pricing which would put telephone services out of reach for many subscribers. Effectively, the inefficiency of the monopoly is used as the killer argument to ensure that the monopoly should stay in place.

If this framework is established as a given, proposals for competition lead to all sorts of difficulties. If competitors are allowed free reign, it is said, they would simply be able to 'cream skim' the profitable customer segments currently being levied to support the local network subsidy. It would be easy for them to win away the lucrative business customers, leaving the incumbent telco with the unprofitable bit of the business – its ability to fund universal service would be gone.

From a market forces stance the solution seemed fairly simple. Stiff competition would force the incumbent to find ways to balance the anomaly. Cutting the costs associated with local network would be a good place to start.

In the UK it was decided a compromise should be reached which diminished the possibility of BT suffering some calamitous financial failure – a political embarrassment which would set the Conservative government's privatisation programme back by 20 years. To hedge the bets it was decided to institute a mechanism called the access deficit. Under this arrangement the duopoly competitor would help fund the supposedly loss-making access network. The money its customers would otherwise have paid BT, through trunk and international call charges to subsidise the access network, would continue to be handed over through the competitor.

BT limbers up

In the early 1980s when telecoms liberalisation and competition were being gingerly introduced, much hand-wringing over the effects of market forces was actually done on behalf of BT itself which, it was thought, was likely to be comprehensively drubbed by lean, commercial competitors.

Telecoms regulation was mostly a way of letting the clutch out slowly, just in case the vehicle went wildly out of control. The role of the regulator as chief clutch operator was supposed to wither away as the market got itself going.

In fact first gear saw BT enjoy the luxury of nearly a decade of low-key competition against Mercury Communications – the duopoly phase. For the most part, Mercury concentrated on the business market. Although BT found Mercury's incursions into its most lucrative market painful, it was left without any serious competition in the residential market and had the breathing space to restructure itself in preparation for the worse competition which was to come.

The duopoly phase was punctuated by minor blips on the radar when BT's growing, or rather shrinking, pains became apparent to customers. Most memorably it found itself in hot water over call box reliability as it struggled to finesse the complicated balancing act between profits, restructuring, union unrest and public pressure applied through the regulator.

BT restructuring

Key to BT's restructuring was the gradual deployment through the 1980s and into the 1990s, of the majority of its digital switching and transmission technology. This enabled it to 'downsize', as the jargon has it. First the huge crop of engineers and operational support staff were pruned as the digital equipment was progressively installed.

Then came Project Sovereign. This saw the company slice its way through a now bloated management structure (there now being far fewer people to manage).

BT has now probably reached an acceptable staff size. But it was forced to prune through a period when its European equivalents remained fairly immune to competitive pressure.

Now, with just over 100,000 employees, down from its late 1980s peak of around 230, 000, BT is a relatively lean operator by European standards.

Without competition it could, almost certainly would, have been very different.

One of the most trenchantly monopolistic of what used to be called the European PTTs, is Deutsche Telekom, BT's German equivalent. Although preparing for both privatisation and a competitive onslaught when the market liberalisation floodgates are opened in Europe in 1998, it still has a gigantic payroll, with an amazing 213,000 employees. By way of comparison, AT&T generates more revenue with just 77,000 employees and this in the desperately competitive US long distance market.

As a competitor, therefore, BT has transformed itself, as cable executives will only too readily admit.

What a difference a decade makes

It is now hard to grasp just how far and how quickly the BT network, and BT's business has developed since privatisation.

Somehow the PR language deployed by the company through the period to describe what it was up to simply didn't evoke the scope of the change. This is partly because BT understandably emphasises 'modernisation', 'digitalisation', 'major investment in infrastructure' and similar abstract terms without dwelling over-much on what was being modernised. At the time of privatisation, much of BT's network was quite simply antiquated – not just slightly out of date but 'antique' in the real sense of that term.

First generation mechanical switches were still in service in even the most prestigious locations. BT's Mayfair exchange for instance, was still operating ancient mechanical equipment in the late 1980s. Ironically, because of Mayfair's status as a location it had a lot of diverts in service from other, in many cases more modern exchanges because of the prestige of the leading number.

Technical progress

Several generations of intermediate technology were installed right through to the 1970s – more compact mechanical switches and then electronic switches with stored program control made their appearance. Even more progress was made with transmission equipment. Frequency division multiplexing, where several voice conversations could share the same physical circuit across the backbone network, were a relatively early development from the 1930s, and time division multiplexing, using pulse code modulation, was also in development from very early on.

But the drive for technological development in both switching and transmission was always muted. On the one side the Post Office, as it then was, was state owned with the result that its capital expenditure budget was always competing with more politically telling expenditure demands, especially in the cash-strapped 1970s. Demand-side pressure was negligible until the 1980s. Voice telephony was a 'stable' application. You picked up the phone, dialled the number, talked and hung up. Nothing could be simpler, nothing could be less in need of major enhancement.

Apart from long distance subscriber dialling (you used to have to be connected by an operator) and the arrival of satellites for international services, it is hard to think of many changes

Against this background it is hardly surprising that major pockets of ancient exchange just kept on keeping on. As bits wore out, they were replaced with new bits.

Revolution

The establishment of the telecoms arm of the cable industry has been an important development, but it is important to remember that for BT and especially its employees, the last decade or so of telecoms liberalisation has seen both structural reform and a complete technical revolution go hand in hand.

For BT's users the results have been profound, but relatively invisible on a day-to-day basis. They still have a telephone (albeit one with buttons) and they tend to pay much less for service and calls than they used to.

The real revolution has taken place deep inside BT's telephone exchanges.

Even seven years ago, many of these anonymous buildings were filled with ancient clanking machinery, covering hundreds of square yards of floor space with dozens of engineers in attendance. Today the same exchange is likely to be doing the same job using a row of fridge-freezer style digital switches and be managed from a computer terminal in another exchange.

Antiques

The clanking machinery was called Strowger exchange equipment. Strowger falls into that rare category of machinery that shows you what it is doing as it is doing it, and it is perhaps no accident that the sort of people who tend to like old telephone exchanges for their own sake (yes, they do exist) often have a weakness for steam engines.

And like steam engines, reasons could always be found to retain them in service well after they should have been thrown on the scrap heap. For a start they were inherently reliable – in computer terms they'd probably be called 'distributed'. If one bit of the Strowger switch broke, it could be repaired while the rest of the switch carried on clicking and whirring around it.

In comparison, today's high technology equipment has an underwhelming physical presence. A demonstration of a state-of-the-art video server pumping scores of individual television signals across a broadband network is likely to materialise as an unassuming little box the size of a hotel room's drinks fridge, parked in the corner of a lab.

To enter a Strowger exchange was to be shrunk to the size of a fingernail and taken on a tour around the insides of a 1930s vintage mechanical adding machine – so little power in such a huge space. But at least you could see exactly what it was doing.

Telecoms tours

Although my tour was about seven years ago now, a host of impressions remain vivid.

For a start, there was the sheer mass of colour-coded twisted pair cable festooned across the ceiling and descending in great swathes to various wiring frames. Then of course there was the ratchet-style clicking of the banks of Strowger selectors, rising, falling and rotating as the calls came in.

But it was not so much the wiring and Strowger kit which held the surprises – this I had seen before – it was the incidental elements. There was what I can only suppose was a circuit tester of some kind. It was like a medium-sized cardboard box mounted on a pair of wheels and pulled about by a handle in much the same way as a golf cart. The enclosure was made of what looked like oak, and the entire contraption seemed designed to yield a test result on a single round needle dial about three inches in diameter.

To be fair, there was also a PC performing some monitoring tasks – a modern life support system just managing to keep the patient from slipping away.

The sense of having been shrunk was more pronounced in the battery room. The battery, which provided the power for all the lines, was about the size of a swimming pool. It was so obviously a huge car battery with its top off – banks of gigantic lead plates immersed in acid.

But the 'new supporting the old' motif was most vivid when we finally came to look at the billing system.

A mechanical exchange times its subscribers' calls by generating inaudible pulses across the open circuit. The pulses step round a counter like an old fashioned electricity meter. When the subscriber hangs up and the circuit is closed, the pulses cease to turn the meter. The pulse rate is increased during peak times, decreased during off-peak, and the difference between readings used to yield each subscriber's usage of 'units' for the quarter.

There had to be one small meter for each subscriber. At my exchange, these were mounted in a large rectangular bank on a board.

Reading the meters, I was told, involved taking a photograph of the board. The image was then blown up and somebody went through taking all the readings from the photographic print.

In the end it was the electronics industry which played the key role in dislodging this static public service model – by both agitating for change so that it could hook its computers up over long distances, and then by providing the digital equipment to make both the advanced services, and competition between them, feasible.

Seeds of discontent

Pressure for change began to build only slowly through the 1960s and 1970s as computer networks came onto the scene and the unstoppable force of the entrepreneurial computer sector began bumping against the immovable object of the state-owned or protected telecoms operator.

Each stage in the history of the computer has been accompanied by an ever-greater data communications requirement. Time-share bureaux allowed horrendously expensive central 'mainframe' computers to be shared by several organisations on a real-time basis (replacing the requirement to send off 'batches' of data in the form of tape or punched cards to be processed as a 'job').

This required data circuits – or at least circuits which could be made to carry data – engineered across the public telephone network. This was the only way to get computer terminals located on company premises to communicate with the central computer at the bureau.

As 'horrendously expensive' gradually turned into 'expensive but feasible', mainframes and the even more affordable minicomputers were cheap enough to be owned by an increasing range of organisations and for an increasing range of applications, but not so cheap that it wasn't still important to share them, even over long distances. So the need to lease permanent telecoms circuits, across which modems could be made to communicate, grew.

Distributed computing

By the end of the 1970s, computer vendors were talking to customers about 'distributed computing'. Instead of a single, central mainframe which did everything at a distance, this alternative approach saw large companies build networks of computers so that more processing could be done at, or nearer the department concerned. While this reduced the dozens of long-distance terminal to host communications channels, it meant that the distributed computers often had to communicate large amounts of critical data amongst themselves. Faster and more reliable circuits were required.

Specialised data communications equipment vendors produced all sorts of products to overcome the limitations of the analogue telecoms network – modems which could use clever encoding techniques to pump more data over a single telecoms circuit; multiplexers which could make use of this extra circuit capacity by concentrating several terminal to host conversations onto a single line; and eventually entire networks of modems and multiplexers which could exchange data between thousands of hosts and terminals all over the globe.

Every step of the way the computer industry felt itself impeded by the bureaucratic sluggishness of monopoly telecoms operators who seemed more intent on preventing than enabling the new.

There was what the industry regarded as predatory pricing for line conditioning and connection, not to mention the line rental pricing itself. And there was always delay. For equipment vendors, laborious, time-consuming and expensive equipment approval procedures threw into stark relief the culture gulf between themselves and the telecoms operators. Where they were concerned with time-to-market, telecoms operators were infuriatingly obsessed with technical and safety issues. Once equipment was approved, there was often more delay before service was provided .. and so on.

Kicking at the foundations

The first cracks appeared in the US. There virtual monopolies, as established by the Bell Telephone company, were at best warily tolerated, unlike Europe where the concept of state owned, public service oriented organisations was accepted almost without demure right across the political spectrum.

US business' demands for a square deal on data services saw the establishment of General Electric Information Services (GEIS) which went on to build a substantial network to support distributed computer applications.

A second major kick at the foundations of the natural monopoly came with what is new celebrated in telecoms as the Carter Phone decision in 1975. This established the customers' rights to attach the equipment of their choice, subject to sensible safeguards of course, to the telephone company's network.

With 'Carterphone' the dogs were set loose. Even 15 years ago, the notion that the future of computers and communications networks were inextricably linked was widely accepted. By the early 1980s the by then booming computer industry was set in stark contrast with the moribund telecoms industry whose activities it was now overlapping at the edges as computers became networked over an ever wider area.

Also by the 1980s, the political winds were blowing in the right direction in the UK and US. In both countries those huge telecoms sectors seemed to exhibit some of the worst aspects of what was viewed as the complacent 'corporatism' of the 1960s and 1970s. They were deemed unresponsive to customers, their services were suspected of being over-priced, and their monopoly status meant they were generally unpopular with the public.

In both countries fresh administrations had been elected on a promise to reinvigorate market forces through supply-side economics and government de-regulation. In both countries the telecoms industry was a sitting target.

In the US the end-result, through judicial process rather than through political initiative, was the break-up of AT&T into local Bell operating companies and a range of pro-competitive measures in the long-distance market.

But the US had approached the issue from a different direction – its reforms were designed to retrieve competition from the sclerosis of monopoly capitalism, and they went with, rather than against, the traditional political grain. Where the US break-up was within its established 'anti-trust' tradition, the liberalisation and privatisation introduced in the UK represented a radical and contentious break with the past.

Major shift

In the UK, it is easy now to forget just how fraught the introduction of liberalisation and competition was. In the early to mid 1980s, the UK's political landscape was still sharply polarised between those who believed that telecommunications, along with a whole range of other utilities and services, could be most coherently and equitably managed in a public service, non-competitive environment, and those who thought the 'market' could be successfully applied to what most people in the telecoms industry itself still regarded as a natural monopoly.

After all, even in the US the local telephone companies were to keep their monopolies over the local loop.

Today the argument seems to have been largely won by the 'marketeers', at least for the time being, and the success of the UK telecoms sector has played a major part in this political shift. But at the time, the introduction of competition and BT's privatisation was more than a radical tweak of an industry sector. They were a key political turning point for the Thatcher government.

Because earlier privatisations had been only partially successful, the government had been determined to make BT's flotation a political as well as a financial triumph. It promoted the offer vigorously, turning it into a celebration of popular capitalism – a carnival event designed to help revolutionise the concept of share ownership in the UK. BT stock immediately traded at a premium and has been rising steadily ever since. The BT float formed the model for the subsequent UK utilities' privatisations – a potent brew of broad share ownership, instant profits on successful share applications, and the promise of price regulation for consumers through the offices of an independent regulator.

Benefits for UK Ltd

But the UK's telecoms revolution has had another, less widely understood, consequence. It not only energised the provision of communications services, but created an open, dynamic market for those supplying the operators with technology and services.

The advantage UK-based telecoms suppliers have gained by 10 to 15 years of operating in a competitive market is profound. Competition subtly changes the way everything – right down to mundane network components – is bought and sold. Not only have these companies developed products and services tailored to the particular requirements of competitive operators, they have inculcated a culture which can respond to further change.

5

Pandora's Boxes

Much is made in the broad computing/communications industry of the idea of convergence – where the boundaries between formerly separate industry sectors are breaking down because digital technology is increasingly underpinning all their activities.

In the communications industry the other side of the same coin is divergence. Digital equipment can now be programmed to do almost anything. So cable operators are now faced by a huge range of service options and their main problem is deciding how to spread finite resources effectively. Do they support basic rate ISDN or cable modems, or both? What about centrex? What about Internet access services?

Before we can discuss these issues we need to outline what all these different services do and how they do them. This chapter is designed to provide a technical overview of some of the current and future service possibilities

Convergence

Digital systems are inherently promiscuous – a stream of data squirted between devices within one sector is just as capable of being squirted over the old sector demarcation lines, enabling one industry to start encroaching on the turf of another.

The convergence of the cable TV and telecoms industry is perhaps the most glaring example of this trend in action, but the same process is at work in a myriad of sectors. For instance once publishing companies are delivering information on-line, perhaps on an hourly basis, they begin to resemble TV channels in terms of the real information or entertainment demand they are meeting. On the other side of the fence, by supplementing TV programmes with fact sheets and Web addresses, TV production companies are beginning to move suspiciously in the direction of publishing.

These individually slight shifts in activity all add up to a gradual realignment of just about every sector of the economy and, as we shall see, the pace will quicken with the increasing take-up of on-line access through the Internet.

At root, the collective shift has come about by a series of deeply pragmatic individual decisions. Digital systems, such as those now employed in telecommunications, were developed because they had a whole range of immediate commercial advantages – they were more reliable, cheaper to manufacture, easier to manage, took up less space, dissipated less heat and so on. They were just better at doing the job of the mechanical thing they were replacing.

But breaking down a broad range of real-world activities (like converting the paper and typewriter into the keyboard and screen I'm now using to write this) and representing it all in numbers has had the effect of throwing everything up into the air.

If everything can be represented by streams of numbers (moving and still pictures, three dimensional objects, music, voice, texture, perhaps even smell) then conversely streams of numbers can be made to represent anything – well beyond the digitalisation of existing services. If you control a network capable of pumping those numbers all over the country, you're facing a startling set of possibilities. This chapter is about the challenges and opportunities posed by that simple equation.

Confusing options

The cable industry faces a confusing array of options in terms of the types of services it is capable of deploying. Its key challenge is to mesh the do-able with the marketable.

At least the main competitor is facing the same problem. Incumbent telcos like BT have long been anxious to develop their networks and services in such a way that as much 'value-added' service as possible can be provided on their side of the public/private network divide. In data communications the track record in this regard has not been good.

In the late 1980s it was common for incumbent telcos to enthuse over their abilities to provide 'value added' once a digital service called ISDN was brought into the front line (see next page). The trenches would move forward and all sorts of private network functionality would find itself behind telco lines where it would provide value-added business for telecoms operators.

Nothing of the sort has occurred and ten years on telcos are still struggling to do things to data apart from just transporting it from A to B. If anything, the front line has moved the other way with more value-added functionality being deployed in private – or at least third party – domains. The rise of the Internet has been another blow, having seen off the telco-dominated X.400 email and X.500 distributed mail addressing scheme, and becoming the focus for networked multimedia development.

With voice, the picture is different. Here it can be argued that many bright spots have emerged, with DDI (Direct Dial In), Centrex, (see page 59) voicemail and enhanced services rolling back part of the front line by successfully transplanting telephone system functions (like divert, hold and so on) into the public network infrastructure, while freephone and premium services have created major new revenue streams.

That telcos have not yet found many ways to add value to data applications in the same way has much to do with the relatively slow development of broadband access, but it also has a lot to do with the historical divergence and complexity of users' networked data applications.

ISDN: Channelling Digits

Developed in the 1970s, the Integrated Services Digital Network (ISDN) concept was the telcos' first response to the new service integration possibilities presented by their planned digital networks.

ISDN was a way of extending the 64,000 digital bits per second data channels, used to move voice about the core network, straight out to the customer instead of converting them back into analogue channels at the local exchange for the final leg of their journey to the home or business. The idea was that special terminals would be made available to customers to allow them to integrate the different types of communications they wished to undertake in the future – videophone, computer terminal, voice, fax. The ISDN service would provide a common means of connecting these different classes of service across the network.

ISDN primary rate services – or DASS circuits as they are usually called in the cable industry because they conform to BT's Digital Access Signalling System – are the bread and butter of the cable industry's offering to medium to large businesses, providing 30 exchange lines across one 2Mbit/s circuit. A DASS circuit is usually attached to the company switchboard (often called a Private Branch Exchange or PBX) at a ratio of around 4 to 5 telephone extension users to one exchange line.

Now several cable Multiple Systems Operators (MSOs) are offering or preparing to offer Basic rate ISDN, (BR ISDN) which was designed to provide two 64kbit/s channels across a standard telephone line.

Cable operators were previously reluctant to offer BR ISDN because it made an ungainly fit with the 'architecture' of their telecoms networks which they had designed to run alongside their cable networks.

Cable networks are usually built around 500 home nodes. Fibre is run to each node – best thought of as a sophisticated junction box. Here the TV signals are converted from the fibre to run across coaxial copper cable for the final few hundred metres to the home.

The telephone network does more or less the same thing using a separate set of cables and connecting equipment. Fibre is usually used to connect the nodes to the telecoms switch at the cable operator's 'head end' (where the television signals are also fed into the cable system). At the node, the individual digital voice channels are broken out (or de-multiplexed) from the high-speed fibre and converted into analogue signals to travel across thin, copper cable to each subscriber's telephone.

BR ISDN however, is specially designed for telecoms networks where the copper runs all the way to a local exchange serving thousands of homes. So cable operators have ended up with a network design which makes it comparatively expensive to deploy services at low density. If only one customer asks for BR ISDN from a 500 home node (a not unlikely occurrence) then the cable operator must expensively upgrade the node and run the service at a loss.

In most cases of course, BR ISDN demand is likely to be clustered in business areas where the possibility of connecting other customers is higher.

But there is no doubt that cable operators have been forced to provide BR ISDN because business customers are asking for it and are reluctant to move from BT if it is not on offer.

Terminal decline

In the late 1970s, early 1980s, the services in line for the sort of integration that ISDN might offer looked quite a bit different. Computer network architectures were still terminal-to-host oriented. A relatively 'dumb' terminal logged into, and conducted a session with, a large mainframe or mini computer. What was required then, and was envisaged for the future, was network support for dial-up sessions. Faster, but essentially the same.

At this time facsimile machines were beginning to be sold in numbers – a faster, digital switched service for fax seemed sensible as well.

The idea (if not the practical reality) of the videophone was also well entrenched so it seemed to make sense at the time that the telephone of the future would have a screen as well and would require a high-speed digital channel to provide the picture.

With all this as background, the telcos were beginning to think of how they could support a set of next-generation digital network applications in such a way that they could all be folded into one set of public network technologies.

ISDN was the result. As was usual, the operators viewed the world from the middle of the public network. ISDN was a way of extending the 64kbit/s channels used in the core telecoms network for transporting voice. Instead of converting the digital streams into traditional analogue signals (and vice versa, of course) for their trip across the final 'local loop' which links individual residential or business lines to the local exchange, ISDN would run the digital channel right to the user's terminal. This meant that the terminal (telephone, PC, videophone) had to be able to do some processing, but it also meant that data devices, like PCs for instance, would no longer have to dial up a destination across an analogue line using a slow, unreliable modem.

Styles and sizes

ISDN has been designed to come in two sizes. Basic Rate ISDN, known in BT parlance as ISDN 2, is the single line version designed to employ the existing twisted pair telecoms cables.

It is essentially a high-speed modem link between the exchange and the user's premises. This arrangement can make the simple copper cable carry two 64kbit/s channels and a 16Kbit/s signalling channel. The original idea was that two channels would be enough simultaneously to support two applications at the terminal – say voice and high speed fax or email at the same time.

Primary rate involves pumping a 2Mbit/s circuit at the end-user, time divided to support 30 64kbit/s channels and a 64kbit/s signalling channel. In the UK, this service is usually defined as a DASS (Digital Access Signalling System). DASS was BT's own primary rate signalling standard and is widely supported on users' PBX equipment.

As things have now evolved, primary rate or DASS circuits are the most efficient way to connect multi-line customers' PBXs to the network, and are used mostly for this purpose. There is a theoretical option available to 'integrate services', but for the most part PBX users tend to equip extension users with relatively simple analogue tone phones. In other words, the digital part of ISDN goes only as far as a company's PBX.

Inflating requirements

In the early 1980s a 64kbit/s channel seemed quite impressive. In fact it was difficult to envisage ever wanting anything faster. But by the time ISDN really started being deployed commercially many applications required faster connections. One technique, called 'inverse multiplexing' is designed to turn a series of channels back into a single pipe (industry terminology for single a conduit).

During the 1980s, large companies had taken advantage of increasingly low priced 2 megabit line services (often called Megastreams, BT's brand name), to build large private voice and data networks to link their offices.

Many of these companies run business-critical applications across these networks. So critical, that if a link on their own network fails, their entire business grinds to a halt.

So it is not unusual for large corporate networks to have duplicate 2Mbit/s pipes. If a critical link fails, its duplicate can go into action without anyone noticing the difference. But even the largest company can't help noticing the huge cost of 2Mbit/s circuit leased just to sit idle in case its mate breaks.

Inverse multiplexers were designed to turn a DASS circuit into a straight 2Mbit/s pipe, so that it can handle things like computer file transfers and the like. However dialling 30 connections at standard BT rates is a very expensive way to establish a 2Mbit/s connection. But because inverse multiplexed connections would only be established when a line failed, or perhaps when existing capacity went into overload, it made good economic sense.

Inverse multiplexing is a little more complicated than it sounds because although a DASS is a 2Mbit/s circuit, it is designed to deal with 64Kbit/s channels. Each channel is demuliplexed at the local exchange and switched individually across the network. It is not possible to have the user equipment dial 30 lines to the same destination and automatically open a 2Mbit/s pipe because each channel can take a separate route across the network and will therefore get itself out of synchronisation with the data travelling on the other 64kbit/s circuits. The inverse multiplexer uses clever technology to make sure a big stream of data goes out and arrives in the right order when it spreads itself across several channels.

For a period, leased line back-up (BR ISDN could be used to back up kilostreams – 64kbit/s leased circuits) was about the most popular BR ISDN use around – a far cry from the original ISDN conception. For a period also this was a real problem for BT. There it was, having invested millions on ISDN development and equipment, while a good proportion of the lines deployed were garnering pitiful line revenue because they were hardly ever being used. But now the growth in demand for fast Internet access and file transfer applications has moved the service on to more profitable (for the operator) uses.

New niche

ISDN now seems to have found a niche, but its initial conception has not been fulfiled. It is not often an integrator of services, just a way of providing 64kbit/s switched digital channels.

It is now widely acknowledged within the telecoms industry that ISDN's failure as a conception marked the end of an era. ISDN was put together as public network solution to a complex and fast changing set of user requirements, but with little reference to customers' actual requirements.

In attempting to bite off so much, ISDN ended up missing its target. As currently the only way of making relatively high speed dial-up connections, it is hardly surprising that it is being taken up by users – but this is not really what it was originally designed to do.

Next generation

One of the most important technical trends in the telecoms business is the move from the industry's 1970s-based digital channel switching technology, represented by ISDN, to the so-called 'broadband' technologies which are expected to replace it.

This section will also give us a chance to describe briefly the technical landscape – hopefully without being too technical.

Telecoms has always had two components – transmission and switching. Transmission gets as much voice signal as possible onto as few circuits as possible and shunts it between switches. The switches switch the signals from a voice channel on one side of the switch to any channel on the other side. Up until recently, apart from a few complications like billing, that was about all the network had to do.

The next generation of broadband networks is designed to handle different types of traffic within one scheme to rationalise the management of the network, decrease costs and reduce the number of engineers employed to maintain it. This has meant that the network technology has had to get a bit more complicated.

Transmission

Next generation digital transmission, fast becoming this generation transmission, is called synchronous digital hierarchy (SDH), and is designed to look after very high speed circuits in a much more management-friendly way than the existing PDH (plesiochronous digital hierarchy). Where PDH synchronises the arrival and departure of the 'bits' on a link-by-link basis, SDH minimises transit delay – the time taken for a bit to travel across the network – by arranging synchronised timing for the whole network. Individual digital channels can therefore easily be extracted from one hierarchy and slotted into another.

SDH is now proving to be the cable telephony operator's flexible friend as the range of applications that cable operators are now tackling grows. As a means of handling transmission in the backbone (the high speed trunk network) it is proving its worth. But as its costs fall (like-for-like, the technology now costs close to old style PDH equipment) it is increasingly being employed in the access network as well.

SDH was designed to manage the huge transmission capacity made available by fibre-optics and microwave technologies in the public network.

It needs to be distinguished from:

Plesiochronous Digital Hierarchy which provides the 64kbit/s and 2.048Mbit/s circuits which are today's common telecoms currency. 'Plesiochronous' means in practice that different transmission links are almost synchronised – or at least that they get out of phase in a way gradual enough to allow the equipment to adjust itself as it goes along.

An asynchronous system is designed in such a way that timing doesn't matter, a plesiochronous system tries to keep everything synchronised but has enough flexibility to cope with the fact that it isn't.. quite.

Sonet

This is the North American standard version of SDH, and the nearly identical technology is often referred to as Sonet/SDH. Europe and most of the rest of the world uses SDH, whereas North America uses Sonet. The differences are in the detail, not the concept, and flow mostly from North America having a different first generation digital hierarchy which Sonet must incorporate.

So it is faster.. what else?

From the telecoms operator's perspective, SDH effectively 'packages' high-speed transmission so that it is easier to handle on a day-to-day basis –

as the CD ROM is to the long playing record, so SDH is to PDH.

To this end it has several important features.

Because of the way the hierarchy is structured, the operator gets more control over individual channels, right down to 2Mbit/s – high speed PDH multiplexers can be difficult to manage and are inherently less flexible

At the fibre level, SDH uses two counter-rotating rings arranged in such a way that any break can be by-passed. One ring remains unused until the fibres are broken.

The nodes on either side of the break each create an instant connection between the two fibres, thus forming a new ring. This is essential because the logic of SDH in the backbone is to replace a mesh of high order PDH pipes.

If a single PDH circuit in a group is lost, its load can be spread amongst the others. SDH has to compensate for its lack of circuit diversity by being inherently reliable through this redundancy.

SDH not only copes with the bandwidth requirements of today, but is designed for a world where vast increases in bandwidth demand are expected for data applications. The idea is that the technology will push its way out towards the business and maybe eventually the home as this demand builds.

Multiplexing

This technique has always been the basis of transmission technology. From quite early this century telecoms operators were already installing frequency division multiplexers to get more out of existing analogue circuits.

By the 1970s it was possible to encode voice patterns in digital format using pulse code modulation (PCM), so time division multiplexing could be widely introduced.

If you encode voice as a stream of binary digits requiring 64kbit/s to provide an adequate sampling rate, by driving a circuit at 128kbit/s, it is possible run two voice channels together by interleaving the bits – the circuit sends one bit for one conversation and then one for the other, just as though it were dealing two hands of cards.

Drive a circuit at 2Mbit/s and the same technique will give you 32 channels – the basis of a primary rate ISDN circuit which has 30 voice channels, a signalling channel and a channel which contains control information.

The digital hierarchy describes the framework within which these collections of voice-sized (64kbit/s) data channels are multiplexed together – first into primary rate (2.048Mbit/s), then four of these are streamed together to provide the next order in the hierarchy at just over 8 Mbit/s.

Third order takes four of the above 'second-order' tributaries and multiplexes these over a link at 34 Mbit/s; fourth order steps this up to 140Mbit/s. The top end of the PDH structure is the fifth order, handling 565 Mbit/s making up 7680 channels.

SDH is designed to come in and take over at the top end of PDH, and goes on to establish a hierarchy through 2.5 gigabits and above.

What's so important about synchronisation?

PDH multiplexers have to jump through a lot of hoops to keep their clock-speeds relatively in tune with the other multiplexers.

This gets more difficult the larger (faster) the underlying circuits become. Think about the amount of data arriving and departing from an 8Mbit/s second order multiplexer. Four 2Mbit/s streams are being interleaved onto a single stream – if data arrives from one of the streams even a tiny bit faster than the average of the other three streams, it won't take long for the system to get out of kilter.

SDH synchronises all the devices on the ring to a single clock to get rid of these tiny timing differences, so that it doesn't have to cope with buffering or 'stuffing' bits all the time.

The grand plan

If fibre has introduced inflation then SDH is designed to provide the higher denomination notes. SDH is designed both to incorporate and perhaps eventually replace PDH as this bandwidth inflation reaches the end user, and applications demand vast pipes.

It seems unlikely, I know, that most business users will ever want multi-megabit services, but history seems to show that applications will always use whatever bandwidth is available, so it is safe to assume that eventually they will.

ATM: Super-fast digits

The conventional digital switching scheme is limited to setting up 64 kilobit per second channels and maintaining them for the duration of a call or session, whether or not meaningful data is flowing. Asynchronous Transfer Mode (ATM) fits nicely on top of SDH transmission and is a way of optimising the network's traffic throughput by allocating variable amounts of bandwidth to the individual applications – my telephone call, your file transfer – as and when it is need.

ATM theoretically liberates applications from their fixed capacity straight jacket. By packing data into individually addressed cells and sending it off across the network it is possible to meet the demands of what are called 'bursty' applications like those generated between a PC and distant fileserver. Here great stretches of time pass during which the devices just exchange a trickle of data across the network to support the network protocols which let one end know that the other is still there. But every so often, the application will generate a 'burst' of data when a file is exchanged. ATM can be configured to handle these bursts.

Reducing delay

ATM also has to support voice and video and that means removing the delay normally associated with addressed packets. ATM is an attempt to get the best of both worlds. Instead of variable length 'packets' of the sort generated across an Ethernet local area network (LAN) or a packet switching network, ATM deals in short fixed length cells of 53 bytes (or Octets in telecoms jargon), five bytes of which are reserved as a 'header' to describe the data and its destination. Because these cells may be travelling at gigabit speeds across a broadband circuit, an entire 53 bytes can actually present itself at a switch interface in less time than it takes a single bit to be presented from a 64kbit/s circuit. So meaningful delay is prevented.

But the really clever part of ATM is the actual switching. Conventional digital switching involves a 'time division' technique where a processor divides its attention between ports to move bits back and forth to accomplish the circuit switching function. ATM was designed for what is called space division switching. At the switch, cells enter a switching matrix where a temporarily-assigned header guides them through a series of binary, left/right decisions. These decision points inside the switch are policed by autonomous, distributed processors. Cells emerge from the matrix at the correct outgoing port without suffering significant delay.

Prospects

So when and under what circumstances does the new infrastructure get deployed? SDH is now being used by telecoms operators, including cable operators, on a wide scale. But public ATM networks have taken longer to develop than at first expected.

ATM, although conceived earlier, started to get really serious in the late 1980s as telecoms infrastructure vendors and telecoms operators began to face up to the theoretical problems involved in switching data running across high speed networks. By 1992, after the establishment of the ATM Forum, the focus switched to user equipment as local network vendors fastened onto

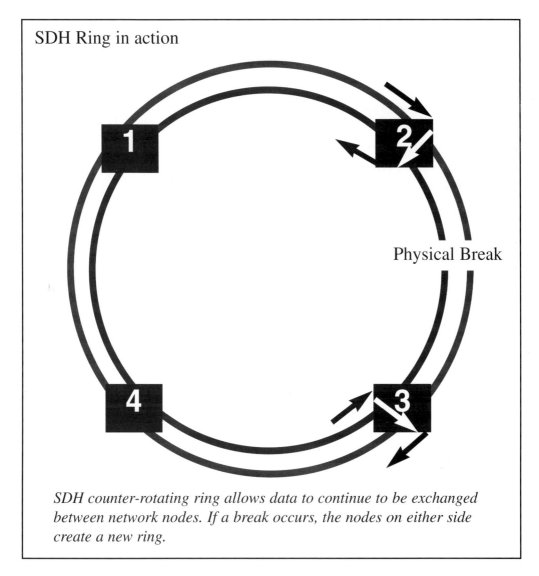

SDH Ring in action

Physical Break

SDH counter-rotating ring allows data to continue to be exchanged between network nodes. If a break occurs, the nodes on either side create a new ring.

ATM as a silver bullet for looming data switching bottlenecks in LANs.

It is probably fair to say that neither the CPE effort or the public network deployments have turned out as expected. On the LAN side, although ATM is developing steadily it has not accomplished the take-over expected four years ago. Two things were supposed to happen which didn't. LAN applications demanding deterministic data delivery (where the data turns up in a regular way) were supposed to come along and be a prime user concern. In fact, there was only desktop videoconferencing and that hasn't gone anywhere.

More importantly bottlenecks and LAN management problems have not proved sufficiently pressing for users to stampede to new, expensive technology. In the meantime, market-driven equipment vendors have found more tactical ways of getting more out of the old technologies – like developing Ethernet switching hubs and now very high speed 2.5 gigabit (2.5 billion bits per second) Ethernet is also on the way.

It has been a similar story on the public network. So far the applications that users say they want supported across the network seldom require the sort of performance offered by a full-scale ATM service. The big infrastructure vendors (the Ericssons and Alcatels) have lately been turning their development effort away from the huge core switches that were originally the envisaged requirement, and are concentrating their fire instead on so-called edge switches, which take feeds from other networks. At this point, ATM is being deployed in specific high speed networks, rather than being used as a wholesale replacement to existing equipment.

More control

Broadband networks don't just involve capacity. Telecoms operators expect a range of benefits from the next generation of network technology.

In an ideal world it must allow the network operator to 'integrate' all services within a single 'full service network' environment. It must also offer significant 'cost of ownership' advantages – it must be self-configuring and self-healing should a link break, and the elements must be remotely configurable to reduce the costs of connection. In effect, operators want a network which looks after itself as much as possible and is always ready to respond to customer demands with a minimum of human intervention. With some luck, and a lot of money, such an advanced, intelligent network will simultaneously be more reliable and easier to manage (thus lowering the operator's costs) and will allow faster and more flexible response to customers' demands.

Lastly, of course, the new technologies have to be 'broad' – capable of delivering vast amounts of data when and where needed for new applications as they develop.

Frame Relay

Frame Relay is a way of organising wide area data applications and will be increasingly supported by cable operators. They can already provide circuits for private Frame Relay networks, but as the cable network build continues and covers the majority of business centres, they will eventually be in a position to provide a national virtual private network (VPN) service for businesses. At present VPNs usually present themselves to users as a

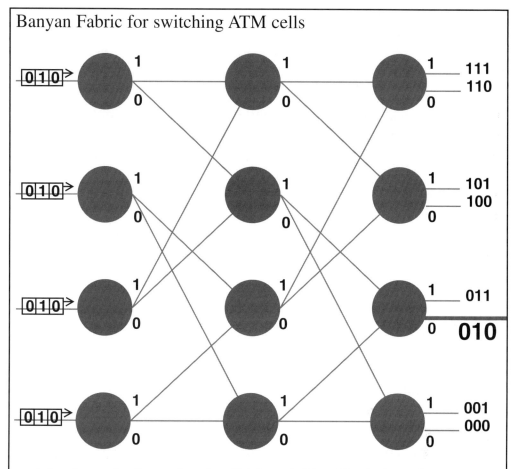

Banyan Fabric for switching ATM cells

A header at the front of each cell (above left) guides it through a sequence of binary gates to emerge at the correct output port – regardless of the entry port. In the above example you can plot the route a cell appearing at any input port would take to get to output port 010.

Frame Relay interface.

As its name suggests, Frame Relay concentrates on relaying variable-sized frames of data from one network to another, regardless of the underlying technologies or users' applications and network operating system.

Unlike X.25-based packet switching from which it is derived, it does not attempt to correct errors in the frames, but relies on the customers' higher level protocols (those operating within the user's network operating system) to validate the data. In practice, all-digital networks have such low bit error rates that the extra delay introduced by the occasional end-system to end-system retransmission is more than compensated for by the elimination of the constant delay generated by the error-checking

Similarly, from the operator's perspective, a frame relay service can be transported across any underlying network, such as an asynchronous transfer mode (ATM) network, for instance. Or it can be delivered across standard digital telecoms circuits.

Both customer and operator can regard frame relay as an interface between the public and private networks. The customer simply presents data to his side of the interface and expects it to appear at the far part of his network as though the service were a piece of cable.

Most services have a Committed Information Rate (CIR) which is a sort of guaranteed delivery level up to a set number of kbit/s. When the customer transmits within the CIR, data must be delivered within an agreed maximum delay by the service provider. Customers are also able to take advantage of the Excess Information Rate (EIR) which allows them to 'burst' data at the port, but without any guarantees on throughput.

Frame Relay services therefore make a nice fit with customer's real application needs – 'bursty' LAN traffic can be catered for as well as time-sensitive, but less data hungry, 'terminal to host' traffic.

The public network operator is also liberated from having to support a proscribed 'top-to-bottom' data network. As his only requirement is to move the data between customers' interfaces and present it properly, he is able to use a flexible mix of network technologies to achieve the service and can upgrade without constraint as needs or opportunities change.

Centrex: Switching private digits

Centrex may prove to be one of the most powerful applications for cable companies in the business market.

Centrex takes advantage of the intelligent capabilities of operators' digital switches by offering customers a 'virtual' switchboard (PBX). Each extension at a customer site is simply multiplexed back to the operator's switching centre where the operator's switch effectively does the PBX's job by providing switching and enhanced features.

The key to Centrex is the concept of the 'group'. Physically, Centrex simply appears as a collection of single lines, but conceptually each line belongs to a single Centrex group with tailored billing and access to enhanced functions. All calls to other members of the group are accomplished through short codes and are unbilled. The service is usually leased at a flat rate over a term and 'off net' calls are billed in the usual way.

Competitive networks tend to have plenty of spare bandwidth capacity as they build market share, so there are no issues surrounding the amount of bandwidth used by tromboning a call all the way to a switching centre and back to the user's premises again when an internal call is made.

For customers, Centrex as a PBX replacement offers a lot of business advantages. It can be designed to have all the standard PBX features – call divert, conference or call back when free, and may also be packaged as a multi-site services, so that companies can throw out leased lines between premises or save money on public switched calls.

Companies don't have to invest in expensive PBXs where they can be trapped by changing needs and fluctuating employee counts.

Although PBXs are inherently reliable, central switches are even more resilient – an added attraction for businesses which live or die by the telephone.

Most PBXs are theoretically 'blocking' but Centrex can allow every employee in a group to be on the phone at the same time.

As currently deployed, there is often a branded split between single and multi-site Centrex – this has more to do with marketing considerations than any technical barriers. While Centrex has been available in the UK since the mid 1980s, it has taken the arrival of meaningful competition between operators to see a real head of steam behind its promotion.

As first deployed, the service was tariffed to meet specific needs on the margin rather than to displace existing lines and premises equipment (usually both supplied by BT). Now, of course, such calculations of relative advantage for the operator are no longer possible. Competition has forced Centrex into the front line. And at the front line if you don't provide it, a competitor will.

The Centrex service offering has a lot of potential for the cable industry. As a competitive tool it makes use of a cable network's natural strength – the end-to-end control it exercises over the local network. Because it can be designed to offer both cost savings and high level service elements, it is seen by many in the industry as a real customer winner. We discuss Centrex as a competitive tool for cable in Chapter 7.

6

The Referee and the Match

The cable industry is regulated from two directions. The ITC, (Independent Television Commission), covers the television side of things, while the Office of Telecommunications (Oftel), the UK's telecoms regulator covers telecoms.

The telecoms regulator's stated goal is to foster the development of sustainable competition in the UK telecoms market. For this simple reason his role has been relatively benign from the cable viewpoint. As cable operators are non-dominant players in a market well and truly dominated by BT, the regulator is most often an ally in the struggle.

As the regulatory regime currently stands, cable operators are exempted from what BT understandably regards as onerous restraints on its competitive behaviour. They are allowed to package (potentially cross-subsidise) services where BT must observe strict accounting separation and a prohibition on unfair cross-subsidisation. They are allowed to discriminate on pricing to different customers, whereas BT can be challenged on its behaviour.

Perhaps most importantly, cable operators are not obliged to provide equal access to other operators, whereas BT must.

These regulatory restraints are all designed to prevent BT using its advantages of scale and scope to crush emerging competition and many of them are likely to be lifted as and when its dominant position is eroded and the competitors are deemed able to stand on their own feet. Indeed, it is always possible that if and when the cable industry begins to approach a 25% market share, it will find versions of these regulatory restraints applied to its own behaviour to protect smaller and newer operators.

Regulation too slow

But despite these regulatory crutches, difficulties remain. The main problem with detailed, prohibitive regulation of the sort applied to BT, is that it makes it impossible for the regulator to take immediate action if the behaviour in question is not specifically prohibited. With the market and its underlying technologies and services changing so quickly, detailed regulation is in a constant state of 'catch-up'.

The current Director General of Telecommunications, Don Cruickshank, maintains he wants Oftel to become a competition authority, rather than a detailed regulator.

He has argued that with competition developing fast in most segments in the telecoms business, the days of tinkering with BT's licence should come to an end and market forces should be brought more into play. At present Oftel and BT must agree to changes to BT's licence. If agreement cannot be reached Oftel can only refer the matter to the UK's Monopolies and Mergers Commission (MMC) – a protracted procedure.

So to ensure that the market forces work smoothly, and corrections can be made in a timely manner, Cruickshank has proposed the development of a new licence condition for all operators which would prohibit anti-competitive behaviour.

Just as importantly, he wants a 'yellow card' enforcement process which would allow him to act quickly and flexibly to correct any misbehaviour.

BT bites watchdog

Cruickshank's original proposal, announced in late 1995, met with a furious response. BT argued that it required a degree of "certainty and due process" if it was to plan its investment on a global basis. If an anti-competitive prohibition were to be established, it wanted an appeals option to go with it. Oftel subsequently agreed to a non-binding appeals panel where BT (or other operators) could further argue their cases.

At the time of writing, BT had agreed to Oftel's amendment with the caveat that BT should be allowed to challenge the validity of the proposed changes in the High Court.

Number portability

One of the protracted wrangles which originally lead Oftel down the road to seeking more flexible powers was over the crucial issue of number portability. The cable companies had long argued that arrangements should be established between competitive operators so that new (cable) customers could take their old (BT) numbers with them when they changed over. The inescapable requirement of having to change numbers was a major customer objection to changing services, especially for businesses. The cable companies argued that lack of number portability was significantly inhibiting competition.

But BT was able to delay change by refusing a licence amendment and forcing a referral to the Monopolies and Mergers Commission (MMC), thus slowing down the introduction of the service.

Long-running dispute

In fact, it was BT's reluctance to offer what Oftel and the cable companies regarded as equitable terms for number portability that led to the breakdown in the Oftel/BT relationship.

Under the current regulatory arrangements, Oftel must agree license modifications with BT. The issue of how portable number costs should be shared was referred to the MMC in April, 1995.

The general principle of portability itself had already been agreed with a modification to BT's licence in 1991. BT was to offer portability to other operators if it were technically feasible, with BT recovering 'reasonable' costs for doing so.

The licence amendment was put to the test when Videotron asked BT for a portability agreement in early 1995. BT wanted Videotron to pay all the costs; Videotron argued that each operator should bear their own; and Oftel thought there should be a middle course with costs being shared.

The MMC eventually determined that the costs of providing number portability should be shared between operators, with the main burden of the costs going to the 'donor' network (in most cases BT).

Oftel followed the MMC guidelines quite closely and determined that the costs split should see 70% being shouldered by BT, 30% by other operators.

The MMC used as its starting point the concept of public interest. It decided that, as portability is in the public interest to the extent that it quickens the pulse of competition and therefore lead to generally lower prices, any cost mechanism which inhibits its introduction must be discouraged.

If BT were simply able to pass all its portability costs on, it would have no incentive to increase its efficiency and lower costs – quite the reverse in fact. In these circumstances a cost-sharing formula which didn't see BT picking up a large proportion of the tab would work against the public interest by inhibiting competition.

Regulatory reasoning

These are the complex lines of reasoning that regulatory bodies use to link their decisions to the concept of public interest. The problem is that this regulatory intervention never ends. In fact as time goes on the regulator tends to create a huge body of precedent that can be picked over and selectively used by other parties in other circumstances.

The end result is that the regulatory process ties itself down in endless argument while the market it is supposed to be regulating changes ever faster.

Cruickshank's attempt to get powers to regulate the nature and thrust of competition rather than to try to prohibit specific behaviour is really a device to struggle out of this legalistic straightjacket. To that extent BT is right. Cruickshank wants to have the ability to be arbitrary and inconsistent in the way he applies the rules. As the referee, he wants the players to accept his decisions on the spot and get on with the game.

Portability in practice

The costs of portability to the donor (BT) involve the system set-up costs (incurred when providing the basic technology to make portability possible between networks); administration (doing the paperwork and making the technical changes required for each customer port) and additional conveyance costs (diverting calls across 'donor' network).

The MCC felt that both administration and conveyancing costs should diminish over time as the technology is applied to them. The extra costs that the cable companies must therefore pay to BT for ported calls are expected to be brought down as this technology has its effect.

Amongst all the tedious technical detail surrounding the portability argument, one issue illustrates the way technical change tends to undo 'set in stone' cost formulae. One of BT's arguments concerned 'tromboning', which it claimed represented a major number portability cost.

Tromboning occurs when calls heading for a 'ported' BT number have to travel all the way to that number's local exchange to discover that its owner now resides on another network. A circuit then has to be established back to the appropriate BT Digital Main Switching Unit (DMSU) for forwarding to the new network – the call path has effectively had to double back on itself. This tromboning, argued BT, would impose significant extra cost.

Unfortunately for BT, it had spent the last decade developing what it terms an 'intelligent network'. With relatively minor software adjustments, BT can invoke 'call dropback' where the switching units basically signal to each other for a split second during call set-up to establish that the number has been ported. The most appropriate call path is then set up across the BT network to route the call to the new 'host' network, so the actual cost is negligible.

Portability in action

Most cable companies expect to offer service to homes and small businesses first and work their way up the scale. At the time of writing, Nynex has introduced a service for single lines and is poised to launch a multi-line service to win business customers. Telewest was about to offer single line portability.

In fact there were few technical issues surrounding the implementation of portability. The main problems were down to inter-company communication and administration procedures.

Before a customer is ported across to cable, all the obvious customer information must be lodged with BT in advance of the 'cut-over'.

To work out a system, Nynex, Telewest and BT engaged in a joint effort to develop an electronic messaging system which would allow information to be exchanged in advance of line installation. The field engineer can then do the installation and call in, identifying himself and the job by using a PIN number to authorise the port's initialisation.

Other problems also need consideration. Ideally, full number portability would be based around a distributed database shared by all the operators. With this in place, every called number could be looked up as it is dialled and a destination and route chosen for each call. Such functionality would also support other advanced services.

Of course no such system is yet in place, so the technicians are wisely proceeding with care.

Nynex was the first cable company to announce a number portability service. Nynex says it is probably the first competitive service (where numbers are ported between competitive operators) in the world.

Providing a portability service for multi-line business users is fraught with difficulty. Creating a solution guaranteed to work 99.99% of the time (a must when providing telecoms service) is complicated by the current incompatibilities between customers' PBX equipment.

Nynex is charging new customers a one-off £19.95 to take their BT numbers with them when they take its cable telephony service.

According to Nynex the portability option will allow cable operators to increase telephony penetration in the areas served by cable by around 10%.

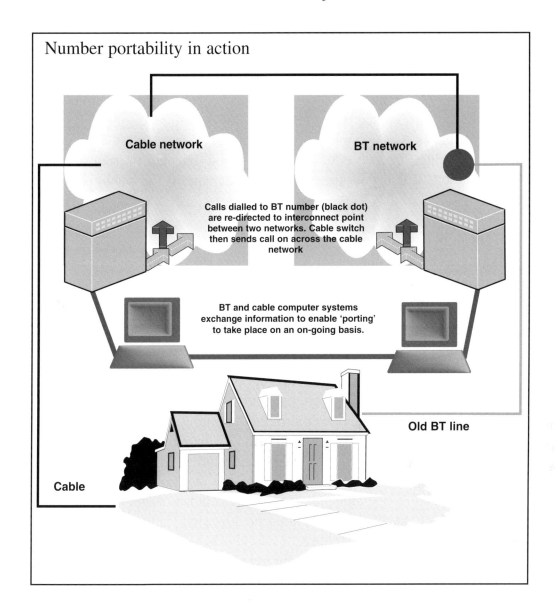

Number portability in action

Cable network

BT network

Calls dialled to BT number (black dot) are re-directed to interconnect point between two networks. Cable switch then sends call on across the cable network

BT and cable computer systems exchange information to enable 'porting' to take place on an on-going basis.

Old BT line

Cable

Customer perception mismatch

Oftel is anxious to aid the pace of competition by acting as a sort of honest broker for comparative performance data, partly as a response to what the cable companies allege is a concerted disinformation effort on cable service by BT.

Cable telecoms services do have to struggle against customer preconceptions about the likely reliability of telephone services running over a cable operator's network.

A study undertaken by Oftel showed that people who use cable telephony rate their operator's customer service level very highly, while existing BT customers often believe cable service to be poor.

In a study of 1100 telephone subscribers, about half of whom were cable customers, 39% of the cable customers rated cable better than BT for telephony, only 18% rated BT better – 43% saw no difference or didn't know.

Clearly this is a communications gap of the highest order. Part of the problem facing the cable companies on the telephony front is their fragmentation, which puts them at a disadvantage in the public relations battle.

The consumer watchdog

The regulator has compiled a set of 'comparable indicators' on operator performance and has published two *Comparable Performance Indicators* booklets, for residential and business customers respectively. They present audited figures on participating operator performance between July and September, 1995, and provide customers with comparative data to help them make an informed choice of operator.

The comparable performance indicators comprise: installing services on time; reliability; fault repair times; complaint handling; and bill accuracy.

On the business side, companies currently providing fewer than 5000 lines, or companies offering service for less than 12 months, are deemed not to have passed Oftel's threshold for reliable indication, and are not included, even though they may be gathering the statistics anyway.

The participating operators for business customers were: BT, Kingston (Hull), and Mercury; with cable represented by Bell Cablemedia, Comcast, General Cable, Nynex and Telewest.

As it turned out, the results were relatively consistent across all the companies – BT, for instance, is usually somewhere in the middle with cable companies both performing better and worse. But, with the odd exception, the spread in performance is not that great. For instance, all companies met their install dates in over 90% of cases and no company had more than five bill accuracy complaints for every 1000 bills issued.

Demonstrating parity

The indicators show a rough parity of performance between leading operators in the UK. Taken as a whole, cable operators do not currently out-perform the competition, but equally the figures provide no basis for any claim from BT that it provides a better service.

Given the different technical and operational conditions experienced by the

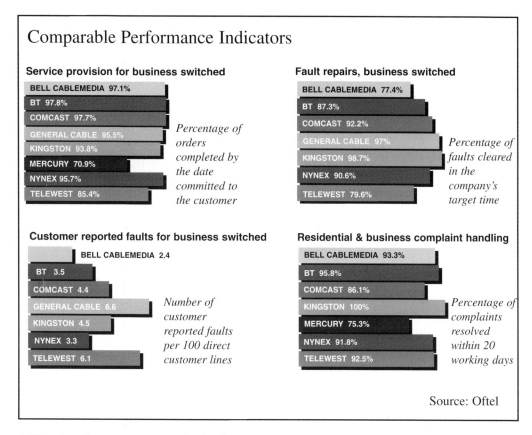

Comparable Performance Indicators

Source: Oftel

UK's leading operators, the indicators may be presented as being remarkably close (see charts, above).

BT's competitive naughtiness

The cable industry has been loud in its denunciation of BT 'dirty tricks'. About the only national coverage it ever gets is when it compiles customer complaints of BT salespeople allegedly misinforming cable customers – or potential cable customers.

In fact it is unlikely that there is a concerted corporate effort by BT to encourage blatant misinformation, simply because it is unlikely to work very well. But BT is inculcating a demanding, results-oriented sales environment, so it is natural that some of its employees have a tendency to over-elaborate a comparative point that they only half understood in the first place.

Some of the more outrageous bits of misinformation alleged have included claims that Oftel didn't 'cover' customers when they moved to a cable company; that a malfunctioning television set attached to cable could 'bring down' a cable customer's telephone service; that cable operators 'tap into' or at least damage BT lines during cable installation; that new cable numbers wouldn't appear in the telephone book; or (a favourite) that cable service meant you could only call other cable subscribers or that cable's local call charges only extended to adjacent streets.

More damaging to cable has been BT's skilled use of discount schemes, which are used selectively to undermine the idea that cable prices are always cheaper. BT's message is that "Switching to Cable doesn't necessarily mean cheaper phone bills." By obfuscating the price comparisons and using a heavy advertising campaign to ram home specific discount schemes, BT has very skillfully increased customer inertia by creating an impression of continuous price cutting.

The little Englander card

BT still thinks it has a winner in the 'foreign ownership' allegation, a point it hammers home in its 'Cable Competitors' fact sheet profiles.

"Most cable companies," says BT, "are massive international organisations, many with headquarters outside Britain."

In fact, much of the finance raised for the development of the cable industry in the UK is raised here. Many operators have shares listed in the UK where customers are able to buy them – just as they're able to buy BT's shares.

And of course, BT is also a "massive international organisation" and a proportion of its shares are no doubt held by 'foreign' investors and pension schemes. In fact its avowed aim is to become a major global telecoms provider. It is especially focused on the US, Germany and Spain, where it has made major investments in competitive operators, directly mirroring the investments made by 'foreign' operators in the UK. It will be interesting to see whether BT thinks it worth continuing this particular line of attack as and when its global ambition really start to bear fruit.

7
The Competitive Squeeze

The introduction of broadband, interactive cable – and the regulatory ability to use it – is proving to be an important element in the global re-invention of the communications and media industries. The digitalisation of both telecoms and cable networks has seen industry sector demarcation lines crumble faster and more profoundly in the communications sector than in perhaps any other.

So far in its history, the UK cable industry has been the beneficiary of technological and regulatory change. Just as the business was beginning to look somewhat lack-lustre in the early 1990s due in part to competition from satellite TV, the further liberalisation of the UK telecoms environment transformed the financial prospects for the industry, and of course, transformed the ownership base as cash-rich foreign telecoms operators saw the opportunities and began buying their way in.

After just five years as full-scale local telecoms operators, most of the cable industry has now seen its telephony revenues outstrip its cable TV revenues and the growth focus for the immediate future is undoubtedly on the telephony sector.

The last two years, in particular, have seen the major cable operators gearing up their business services operations. All have separate business sales divisions and are looking to make major inroads into some key business sectors over the next few years. But others can also benefit from digitalisation and deregulation.

Margins in the telephony business are bound to diminish due to the activities of the regulator on the one hand and market forces on the other. All the participants understand this dynamic well – the dimensions up for argument concern how quickly such erosion will proceed and how its effects on cash flow might be mitigated by developing new, more profitable, strands of businesses.

Slicing the market

Just five years ago, with the ending of the UK's telecoms duopoly, where BT and Mercury were the licensed telecoms operators, the cable companies were the new kids on the block. Their local delivery networks were seen as a vehicle to introduce real infrastructure competition into the UK telephony market and quicken the pulse of competition. Not only do the cable companies now face nimble-footed competition from the incumbent, BT, but also from other new operators.

At one end of the scene are the growing band of what used to be called International Simple Resellers (ISRs) – with recent changes in the telecoms regulations, companies selling international services are able to become fully licensed carriers from the UK. This broad category of competitors includes those with their roots in callback at one end, right through to AT&T which offers both national and international switched services at the other. Virtual Private Network operators – often foreign carriers like France Télécom – also offer alternative services to business customers. And there are the other infrastructure operators, including Mercury, the original BT competitor, and Energis, an electricity company spin-off. These competitors now nibble at the other flank, by competing in private network circuits and 2Mbit/s switched access.

Even the residential market, which cable formerly shared only with BT, is under threat from radio operators. At present this emerging competition remains technologically limited to residential and small business, but radio spectrum capable of delivering primary rate pipes to business customers has already been licensed, so more serious competition may eventually emerge from radio in multi-line access as well.

Segmentation

These diverse players have just one thing in common – they aim to build a market position by concentrating on a single aspect of the telecommunications business and they all, to some extent, illustrate what business theorists might term horizontal segmentation – the tendency of a sector facing severe competition to splinter because the complexity of the business favours specialisation – perhaps even on a global basis – rather than vertical integration.

Telecoms regulation which promotes fair interconnection arrangements and the high technology which allows the interconnection to be accomplished, have both played their part in the process. But there's more to it. The same sort of generalised specialisation is being experienced right across the high technology sector – in electronics, computing and, in perhaps its most advanced form, in the development of businesses which operate across or provide products or services to, the Internet.

The combined effect of the increasing competition has been to make low margin commodities of the bread and butter business – single line and primary rate switched access.

International simple resellers

The UK's open telecoms services market has encouraged the development of businesses providing international services. These organisations are often called International Simple Resellers (ISRs). On the surface, the business is very simple – it involves buying circuit capacity as leased circuits or as wholesale 'call minutes' from other international carriers – and then retailing them direct to customers at a competitive rate and retaining a margin.

But the ISR terminology is rapidly running out of date. While the established telecoms industry likes to characterise these new international telcos as 'cowboys', in the UK at least they have a well-established regulatory and business niche and are now able to obtain licences to become UK-based international carriers.

But the ISRs business image, at least in the industry where they have one, is not helped by many of the companies having clear roots in what is called 'callback' or 'dialback'.

Callback was, and still is in many countries, the original method use to subvert expensive international rates where there existed gross discrepancies between in-bound and out-bound call charges. Such inversions were most likely to occur between the US and countries where a monopoly operator is holding its rates high. AT&T say, operating in a competitive environment, might charge 50 cents a minute for its customers to call a particular country in South America. The South American country's monopoly telco might be charging its customers $2.00 per minute to call the US.

One application for callback was therefore a straightforward one – the customer dialled in, dialled a PIN, the US end dialled back to establish the call – the user then entered the destination US number and was connected at US rates.

But of course it didn't end there. From the same South American country the callback operator could allow the customer to dial a third country from the US – and still be able to make a profit and offer a significant saving to the customer.

Wedge for competition

The battle between callback operators and the telcos in many countries which are still trying to maintain a telecoms monopoly, rages on. In the US, there is a considerable degree of quiet approval of callback operators who are seen to be subverting an unfair international cartel and forcing changes which may eventually lead to liberalisation and more open access for US operators.

The next stage, as it were, for the callback-style operator is to lease its own lines or capacity to provide international service. As we have already seen, this business is often referred to as International Simple Resale (ISR).

Some explanation is called for. In the UK an intermediate phase of telecoms liberalisation established the principle of Value Added Network Services (VANS). Network services which transported data or voice across leased lines were therefore allowed in competition with BT or Mercury if there was deemed to be value added in the process. Straight selling on of capacity where no 'value' had been added was prohibited during this phase and was described as 'simple resale'.

When International 'Simple Resale' was made legitimate therefore, its standard bearers came to be called International Simple Resellers (ISRs) – a term, incidentally, which they detest.

With reason. For many players, the long term business plan is to remain in the market providing all sorts of other international and national services when the margins are eventually squeezed to the point of no return. Like the cable companies operating in the local access market, the new international operators will seek to tailor services for specific niches or specialist applications.

And it is not just international capacity on sale. As the market matures, the new operators are also providing national services by undercutting other companies' trunk rates inside the UK.

With national services, though, the margins are that much tighter.

Cable response

The cable industry is in somewhat of a quandary with international services – it would like its customers to use its own international service, by default as it were. At present there is no regulatory requirement for cable operators, as non-dominant operators, to provide interconnection for their customers to competitive services. On the other hand it seems unwilling to cast itself in the role of anti-competitive demon by becoming embroiled in a controversy over refusing interconnection.

In the meantime, BT has so far decided not to make an issue of it, presumably on the basis that it would hardly be in its interest to draw attention to its own customers' ability to make use of ISR services.

Ironically, with the completion of the latest stage of BT's network modernisation, the mechanics of running these businesses has changed markedly. Instead of cumbersome access codes and PIN numbers, ISR customers can dial (or set the PBX up to dial) a four digit prefix before the dialled number. BT's Calling Line Identification (CLI) allows the ISR to validate the call.

Some ISRs are building a straight-forward business around a cost-saving proposition. The strategy is to target high telecoms spend sectors and build a tactical business by exploiting the arbitrage opportunity between wholesale international circuit rates and the prices charged by the dominant carrier. Simple Resale describes the underlying mechanics of the market opportunity, but some players see it as a jumping-off point rather than an end in itself.

Short-term opportunity

As the market becomes more competitive, the simple cost-saving proposition is thought likely by many observers to lose its gloss as the wholesale/retail margin becomes narrower and customers become less cost-sensitive. Some players are therefore approaching the market with what they claim is a longer-term strategy based around adding 'real value' to the concept.

Once they win customers and gain familiarity with the needs of 'niches' (banks, solicitors, candlestick makers, and so on) they will devise other 'value-added' offerings. These could be anything – calling card services, Internet access, interconnection with cellular services to offer single number personal communications – whatever they determine will appeal to targetable sectors.

One of the most successful companies currently in the UK market is ACC. It launched its service in 1995 and by mid-1996 was carrying 15 million billable minutes of traffic per month. It claims that in July 1996 alone, it won some 4,000 business and residential customers.

ACC claims that it offers savings on business tariffs of about 22%. National daytime calls are between 21% and 24% below BT's basic rate, while daytime calls to Japan are now 28% below BT's basic rate.

Some of the most interesting approaches belong to foreign national carriers, several of which have recently entered the UK market. Backed by large, relatively cash-rich operators, their strategies are predictably longer term and must be viewed within the context of their parent companies' global ambitions.

Nearly all national operators now have 'global ambitions' even if they are essentially defensive.

Target: the small/medium company

One of these carriers is Telia, the dominant Swedish operator. Like the UK, Sweden has a liberal regulatory environment which fosters competition (the OECD claims it is more liberalised than the UK).

Telia itself admits that part of its reason for entering the UK market is to gain experience of operating as a niche competitor. This will help it to counter the considerable competition it is currently experiencing in its home market and learn lessons it can apply to other EU markets towards the end of the decade as they allow competition.

Telia has decided to target small to medium enterprises (SMEs) with relatively large telecoms spends. This is a sector it claims is under-served in the UK market by the majors (BT and Mercury).

So the first problem it faced was how to target its chosen niche. From a marketing perspective, the company claims, there was no easy way to find, let alone convert, this category of user.

The conventional industry sector structure against which most of the UK's direct marketing lists are built does not provide the necessary information. Company size, for instance, is no longer a reliable guide to telecoms spend as some relatively tiny organisations can be participating in a global business – generating a telephone bill out of all proportion to their size.

Similarly, many organisations at the top end of the SME sector may have a very small spend. So the company has embarked on the task of building up its own list of 50,000 target companies from which it hopes to win 25%.

Telia has devised the equivalent of a business telecoms psychometric test which it intends to refine as it learns more about its customers. This way it can target potential clients who are likely to fit the profile, from incomplete data.

Its initial sales proposition involves 30% savings on international calls. As users are fast getting wise to the obscurantist tariffs that are pretty much the norm in this area, Telia is making much of a simplified tariffing approach – an easy-to-compare structure for the main international markets, a uniform

UK trunk call rate, and a range of differentiated prices to low volume destinations.

Ionica

Ionica is just one of several operators expected to provide competition to cable in the local network part of the business by using radio to replace conventional cable as a way of reaching homes and businesses.

Ionica officially launched its radio connection service in mid 1996 and says it will roll out the network on a TV region-by-TV region basis and will be able to offer service to the 'majority' of the UK's population by around mid-1998.

The company claims that it has been financed on the basis that it will achieve a 5% market share of the small business and residential market within five years and says that expressions of interest and market research findings indicate that it may exceed this projection.

This is an ambitious goal. There are currently 21.5 million fixed residential lines in the UK and another 3.5 million small business lines. On this basis Ionica expects to win 1.25 million lines.

Tuning in

Ionica uses a 'fixed radio' system to link its telecommunications service to subscribers. The technology, called Proximity 1 Fixed Radio Access system, is a joint development with Northern Telecom, and will also be used by Scottish Telecom, the telecoms subsidiary of Scottish Power, which plans to use Ionica's frequencies in Scotland to deliver a similar service.

Customers are required to site small antennae at their premises and the operator builds radio transmission base stations across the country to service them. In effect, the system works in a similar way to a cellular network except that the radio links are engineered between fixed locations. The customers use their existing telephone systems but instead of the building wiring being connected to a wireline operator like BT or a cable company, the telephone is linked to the antenna.

Ionica claims that this approach is substantially less capital intensive than deploying the traditional wireline systems of its UK competitors and allows the company to pass these savings on to customers in the form of cheaper call charges.

Under the terms of its licence, Ionica must cover 75% of the UK population within four years of launch, but unlike the cable companies, who must invest substantial sums in 'passing' their franchise customers (nearly 80% of whom have so far not taken up telephone service) the radio access system allows Ionica to more closely match its infrastructure spend with customer take-up as it evolves. On paper this is a considerable competitive advantage.

On the other hand, the narrow-band radio technology which the company seems to be having so much trouble deploying, is best targeted at residential or small business customers with just a handful of lines. It is not suitable for the delivery of high speed data services or the multi-line telephony required by medium to large businesses.

However, Ionica does plan to offer more. Along with Mercury Communications, Scottish Telecom and NTL (see below), it recently won the right to use radio spectrum in the 10 GHz band to deliver broadband data services. Once developed, the technology enabling this generation of radio access will probably support Primary Rate ISDN (2Mbit/s) and perhaps other data transmission standards. Eventual service offerings, however, are almost certainly some years away.

Broadband radio

The cable industry is also expecting more competition from radio networks in providing multi-line and broadband pipes to business customers. As well as Ionica, Mercury, Scottish Power and NTL (the old ITV transmission network, now owned by cable operator Cabletel) have won licenses to offer fixed digital radio access services in the 10 GHz frequency range. The government claims the move will eventually provide more choice of ISDN access for small to medium-sized businesses – at basic rate (144kbit/s) and primary rate (2 Mbit/s).

As far as broadband radio is concerned, it is very early days. Technology has to be both developed and trialed, and networks built or enhanced.

We are not alone

Very few things have excited as much debate in telecoms over the past five years as mobile satellite services (MSS). At first derided as totally impractical when plans were first floated in the early years of the decade, there now looks every chance that they will get into the air. At least one of the schemes envisages providing Internet access. Unlikely as it seems, UK cable may also be facing competition from space.

MSS or LEOs (Low Earth Orbit) as they are often dubbed, require low- or medium-altitude satellite constellations (involving as many as hundreds of individual 'birds') to act as an orbiting cellular telephone system. Conventional satellites adopt a geostationery orbit above the equator, but to match their orbital frequency to 24 hours (so that they stay in the same position) they have to be hoisted a considerable distance into space. Weak signals travelling a vast distance usually require directional dishes to pick them up – communicating directly with hand-held cellular-type terminals – except that the network moves and the user stays (relatively) stationery. There are now a wide variety of MSS in various stages of preparation, but the two deemed most likely to succeed in getting off the ground are Iridium and GlobalStar.

Technology

The technology which supports MSS is diverse. All the systems have been designed to trade off the inherent satellite parameters to meet different service profiles.

In simple terms, the lower the orbit, the lower the transmit power requirement (for both satellite and handset), and the more total capacity. But lower orbits require more satellites and offer a lower worst-case reception angle. Higher orbits mean fewer satellites (less cost and more easily manageable) but less capacity and more transmission power. Higher orbits also introduce circuit delay and are less suitable for voice applications.

Iridium, the first really large project to be announced, looked, with its satellite-to-satellite communications capability, suspiciously like a 'bypass' system to developing country operators. As originally described, it seemed targeted exclusively at affluent businessmen or yachtsmen. At US$2.00 per minute the service not only didn't offer a service benefit to the vast majority of the developing world, but it also looked likely to take lucrative international call revenues away from national telecoms operators.

MSS operators were granted radio spectrum, but national licences to operate telecoms service were still required, so Iridium discovered the idea of dual mode handsets and the importance of involving national operators in the service. GlobalStar has tailored its scheme to national operator sensitivities from day one.

So for both political and economic reasons, the MSS operators' challenge will be to design a diversity of distinct revenue streams each operating at different price-points and exhibiting different service quality and other parameters to justify the differentials. While headline pricing of between $1 to $2.50 per minute is being quoted, it makes sense that spare capacity can be soaked up by offering services to rural areas in developing countries in particular, at much lower rates.

The main contenders.

Iridium differs from some of the other schemes in that its design is literally a flying network, complete with satellite-to-satellite communications. This made Iridium's 66 satellites (formerly 77, hence the name) more expensive and heavier, but Iridium does provide a way for its operator to maximise revenues by using the satellites to trunk the global network.

GlobalStar's scheme takes an opposite tack. Instead of an expensive airborne switching network, it offers the 'bent pipe' transponder model, where signals are simply bounced back to earthstations. This approach is cheaper to deploy and relies upon alliances with existing operators within specific territories to provide the earthstations and the onward call routing. GlobalStar has played the 'global social responsibility' card and is promoting its service as the operator-friendly version which will support and enhance the range of service offerings available to national operators.

Cyberspace

Teledesic is the most ambitious of all the schemes. This is real Cyberspace – a huge constellation of 840 satellites which will, according to the plan (which still appears to be at its early stages), provide data services, rather than voice. Teledesic plans to beat the cost crunch with a mass-production system and rolling launch programme which will continuously replenish the satellite flock (expected to lose dozens of satellites annually).

How does cable compete?

As we can see from the above examples, the cable industry faces a range of actual and future competition, much of it enabled by new technologies. In acquiring its telephony arm, the cable business has found itself building a series of what the telecoms industry now calls access networks. As such it has found itself competing across a range of what, in telephony, are now increasingly seen as distinct businesses in their own right. In this context there is internal debate about where the ultimate 'value' in the business will lie and where the cable companies should concentrate their efforts.

To an extent, this dilemma is at the heart of the whole IT business. Where does the balance of advantage lie? Do you concentrate on doing one thing really well and hope that this will allow you to build a position with and against other players in the industry? Or is there ultimate advantage to be had in building end-to-end ownership over some key functions where you can control the shape of specific services, maybe even make them more attractive than those of your competitors?

The second option sounds attractive but risks spreading relatively limited resources and management focus too thinly. The danger is that the organisation performs only adequately at each value point in an increasingly competitive market and sees its business suffer as a result.

Using the network

The UK's cable network operators are using Centrex to carve out a position in the business market which will prove resistant to competition from 'commodity' providers offering low-margin switched access and leased lines.

For many of the UK's dozen or so substantial Multiple System Operators, Centrex is an important way to capitalise on the advantages of owning an access network. Through it they hope to capitalise on the service functionality other operators can't reach and therefore retain hard-won customers for the long term.

From a strategic cable perspective, building a solid base of Centrex customers therefore makes a lot of sense. With their modern digital switching equipment and plentiful bandwidth in the distribution network, they are all technically able to support the service. Organisationally, too, the level of local customer support required makes Centrex appropriate for cable, while competitors like Energis are better structured for national private network services.

The cable companies see opportunities at both ends of the market. Small to medium-sized businesses (without the size to employ in-house telecoms managers) are prime targets, but so too are large, regionally-based organisations like local authorities. Here they have the coverage to match their needs over several sites.

Those operators with two or more years of serious Centrex experience have generally seen the proportion of Centrex to total business sales climb to healthy proportions. Cambridge Cable, for instance, claims that the majority of its new business sales are now Centrex.

Owning the customer

By its very nature, Centrex works most coherently as a total telecoms solution, especially for a small to medium-sized business customer – relieving a stretched management from the incomprehensible techno-speak and number-crunching required to fine-tune a telecoms system in a competitive environment. Instead of using two or more operators (eg BT lines in, competitive operator's lines out) and having to purchase and manage customer premises equipment, the entire business function can be off-loaded on to a third party.

For the cable operator the Centrex prize involves winning all the lines (otherwise the Centrex functions can't work anyway) and therefore minimising incursions from specialised operators. If a company can be won to Centrex the nature of the beast means that it will probably stay won, and the customer is restricted in his ability to make use of other operators. But cable companies are having to think carefully about how they structure their offerings so that they don't reduce their options further up the line.

As a general rule the approach is to present three tiers. Small Centrex – usually offering somewhere below 50 lines and usually restricted to a single site; multi-site Centrex which is aimed at the high end, and a Centrex offering which enables some degree of interworking with existing PBX equipment. Arrangements are further complicated by customers requiring data circuits as well.

Much as cable operators would like to make Centrex a 'solutions sale', in such a competitive market many customers are naturally going to be expecting price savings. Where multiple sites are involved, cost savings may be a natural feature of a Centrex regime anyway, given that the facility will be replacing switched calls or leased lines between premises. But for single sites the package has to be structured to make it attractive as a simple replacement for an on-site PBX.

As time goes on the businesses and organisations with which Centrex is a natural fit will be either won or lost within cable franchise areas. It will then be necessary to introduce more tiering to attract business at what is now the Centrex margin. This will probably involve stripping out enhanced features for lower leasing costs in the expectation of being able to sell more functionality as time goes on.

One-stop pipe

In the context of the long-term strategic objective to make Centrex the one-stop pipe for local businesses (large and small) the issue of allowing customers to engineer interworking with other customer premises equipment, (and therefore, other services) becomes a real one.

From the cable perspective, the economics of Centrex boil down to making a significant 'up-front' investment in switching and transmission facilities to connect each customer in the expectation of a commensurate return through the period of the contract. The return is made up of predictable Centrex flat-rate lease revenues, but is topped up with less predictable revenue from 'off-net' calls (to non-Centrex group destinations).

As the services market heats up, especially for international calls, allowing customers to connect cable Centrex to on-site equipment means putting the off-net revenue at risk and ruining the economics.

But moving customers to Centrex without the ability to connect it locally cuts off this option. As non-dominant telcos, cable operators are not obliged to allow their customers access to competitive national or international carriers, so for the time-being at least the cable companies are allowed to 'ring-fence' all their switched services. If customers retain a hybrid arrangement however, they can reach competitors through BT (which must provide access).

At present the regulator's policy is to allow cable companies to block competitive access in the interests of promoting infrastructure investment. As and when the cable companies build a 25% share of the market (even if only on a regional basis) this policy can be expected to be reversed.

Customer pressure

But of course pressure for equal access may come from cable customers themselves, especially as licensed international carriers begin stepping up their marketing activities and the situation becomes glaringly obvious. Predictably, therefore, there is a divergence of opinion between the different cable companies on the best way to proceed. Some are holding back from offering hybrid Centrex service (ostensibly because of worries about technical compatibility between PBX and Centrex) until all the ramifications become clear, while others have decided to go with the flow in the interests of meeting customers' requirements.

Ironically, in other circumstances, BT would certainly seize on what could be interpreted as some cable companies underhandedly restricting customers' access to cheaper national and international services. But as this would involve BT highlighting its own customers' ability to make use of alternative carriers, it has presumably decided that discretion is the better part of valour.

Centrex seems tailor-made for cable – at least at this stage in the industry's development. It allows cable business salespeople to 'sell solutions' rather than having to fall back on pure pricing to get a foot in the door. Better yet, the small-to-medium business market, where much of its attention will be concentrated, has historically been under-served by BT. For BT, promoting multi-site Centrex would simply have meant robbing the Peter responsible for leased line and switched revenue, to pay the Paul of its Centrex alternative. Cable salespeople are usually approaching prospects rather than established customers, so there is no revenue downside – each new customer just represents more on the bottom line.

But the real battle may be between business cultures.

Large but local

Despite consolidation into a handful of large groups, the cable companies will remain locally focused and they are likely to continue to enjoy labyrinthine ownership structures. The choice of the word 'enjoy' is not one which would often be used by many within the cable industry: on a day-to-day basis the relative fragmentation of the industry seems to put it at a severe disadvantage when competing with BT in telephony.

The original system for allocating franchises followed a conventional cable TV model – with relatively small territories based on parcels of households. As consolidation has progressed, especially since 1991, what has emerged has been a pattern of patchy ownership on a national level, with the large MSOs in possession of weird geographical scatterings of franchises.

In the cable TV business this level of fragmentation is probably sustainable since, at least in comparison to competitive telecoms, it is a simple utility business which exists well with a local business model.

But with telecoms added to the pot, consolidation has an over-riding logic – and consolidation which generated contiguous blocks of franchises would probably be best of all. Most obviously contiguous franchises allow their MSOs to build their own high-speed links between both their own switches and those of adjoining MSOs or national networks, in the case of Mercury.

In the residential and small business market this only becomes a real advantage when cable companies start carrying enough of the total national switched traffic to make a joint national backbone worthwhile, but such abilities do pay off immediately in the business market. Here blanket coverage of a conurbation allows cable to compete equally on leased circuits or Centrex against national carriers for multi-site but regionally-based businesses or public sector organisations.

Easier to manage

Consolidation, such as that engineered by the establishment of Cable and Wireless Communications, make the management and the deployment of resources easier. At present MSOs may have hundreds of miles between important clusters of franchises. And it also kicks in much needed economies of scale in what is becoming an increasingly diverse business.

With the deployment of services like Centrex and ISDN, not to mention

Internet access and advanced broadband, substantial consolidation would give the remaining MSOs the bulk to build a reasonable range of in-house technical expertise to cover all the bases, and the larger the call minutes handled or cable TV subscribers signed up, the greater the negotiating power.

But most importantly, consolidation into contiguous geographical blocks would allow MSOs to build strong regional identities and market themselves effectively using regional media.

For all these reasons further consolidation is expected, but the process itself has its disadvantages. It means that senior management is engaged in continual negotiation and financial manoeuvring which is obviously a distraction from the day-to-day business of building networks and marketing services.

The power of scope

Unlike the cable industry, BT can leverage its national scope. It is able to deploy coherently national services for large businesses who, in turn, require national coverage to connect sites across the UK. It can mount national marketing efforts – making effective use of national media. In dozens of other ways it has the advantage of scale – on both technical and marketing research, with its buying power, with its global clout, in its ability to form powerful alliances with other telcos or allied businesses.

All of these undoubted advantages have their corollaries. Like all national telephone companies, BT is a slow-moving monolith in a sector which seems, with a few exceptions, to be favouring the nimble and the focused. The problem is that the disadvantages are immediately apparent, but the advantages are glacial.

Certainly, the cable industry is continuing to consolidate, but this is unlikely to result in a single cable company. While the larger MSOs are organising themselves into coherent groups of franchises, many participating franchises remain relatively autonomous affiliates.

But another trend is also apparent.

New service providers

As the access networks organise themselves into geographical entities, the business is also seeing new types of company emerge to tackle horizontal functions – London Interconnect and Northern Interconnect, for instance, operate as separate companies within the federation.Cabletel has acquired a national backbone network with NTL, and MSOs have launched completely separate companies to act as Internet access providers.

It is possible to envisage other initiatives in this area designed to exploit network services possibilities. As build continues, and cable is within splicing distance of most town and city centres and business parks, the cable industry will provide a platform for national Virtual Private Network Services, or, a slight variation of the same idea, for national cross-franchise Centrex services.

However quickly consolidation occurs at the geographical franchise level, the essential genetic structure of the industry will remain intact.

Specialised strata

While the largest cable group comprising Nynex, Bell Cablemedia and Videotron is, at the time of writing, about to merge with national operator Mercury to become Cable and Wireless Communications, autonomous business units will remain an important element for the rest of the cable industry. Cable's strength against the incumbent telco will, ultimately, not be a battle of technologies – much more a battle of two opposing business models. Cable lacks brute strength and a coherent voice, say its critics. But such observations boil down to saying that cable companies don't collectively resemble BT. Cable just might have a real advantage within that organic framework.

In day-to-day terms, it allows a cable operator, especially in the growth phase of service development, to be driven by customers' needs – management jargon for designing and providing specific solutions, or even entire services, when enough people seem to want them. Or even when a single, large business or public sector customer, wants a specific service.

On the business side, especially, this is a real advantage and not simply a transitory one which can only be applied as the first generation of business customers are recruited.

Cable companies are not embryonic BTs and will not lose the ability to be solution-oriented as they consolidate into a handful of large groups.

The days of 'this is our service, take it or leave it', are over for ever. Successful players in a competitive market must continue to present themselves to business customers as service providers not as network operators.

8

The Organic Internet

U ntil recently, the incumbent telecoms sector as a whole still harboured the notion that the industry was experiencing a period of technological and regulatory upheaval which would iron itself out and then life could get back to normal with the old verities and structures in place. Competition and liberalisation were a sort of spasm, an over-correction applied to an industry that had become lethargic. But beneath it all, telecoms was still telecoms. There would be competition, of course, but things would get sensible again.

But it is now clear to nearly everyone that this won't happen. As far as it is possible to see, the only constant in the communications business will be change itself and success will come to companies who can not only cope with change, but actually require its adrenaline just to keep functioning. Coping with constant change means changing your entrenched perspective when you encounter a new approach or a potential threat.

The Internet

The Internet and the underlying service and business model that it is fast coming to represent are the most dramatic harbingers of this new reality. As we have already seen, the investment rationale behind cable hinges on the idea that interactive services will ultimately emerge on the back of high-speed public networks. Until recently it was assumed that these services would be deployed and controlled by those who owned the network. This was a natural enough assumption for the telecoms industry.

But the Internet has released the genie of the 'open' network. The Web is here; it works; and and it is popular with both content providers and users. There is no possibility that the open network genie will go back in the bottle.

It is therefore worth discussing the Internet and the Worldwide Web at length. Not only does Internet access represent an important short term revenue stream for cable, but it also represents an embryonic version of the business model which will develop for high speed, multimedia interactive services.

Creeping up from behind

For many years the telecoms sector simply ignored the Internet. Through the 1980s it backed a range of initiatives which provided telco versions of the same functionality goals as the open Internet-based equivalents – almost without exception, all of these have been consigned to the dustbin, while the open system alternatives have become the *de facto* standards.

With email, the telcos developed and promoted a scheme called X.400 as the standard for the message format. An allied set of standards, called X.500 was supposed to be the basis for a global distributed addressing system. Today we have Internet mail standards and its global addressing system. The proprietary telco-run network of email systems has not materialised.

The Internet's Worldwide Web application has, until recently, been similarly ignored. Despite its huge growth rate, the conventional wisdom was that the Web was a temporary aberration that would subside when 'real' telco-delivered services arrived.

The Internet was architecturally all wrong. It couldn't provide the levels of service that both business and residential customers would soon come to demand.

Wrong again.

Then came the possibility of voice over the Internet. It would never work. If it did, the quality of the voice would be so bad it would prove unacceptable. Now it is apparent that it will work, it will prove acceptable quality-wise, and it does represent a potential threat to international voice revenues. In all cases the 'new' was judged against a set of legacy assumptions about how services could and should be provided.

This chapter outlines the basics of the Internet and how its premier application, the Worldwide Web, has provided a dramatic demonstration of the power of network technology when it is made available in an open and accessible way.

Many in the cable industry believe that providing Internet access services is natural cable territory and will be an important first step towards the development of the all-important third cable service arm. This chapter lays out the basics of the Internet. The following chapters will explore its possibilities for the cable industry.

Probably not interim

It would be a mistake to think of the Internet as a sprawling, self-built, untidy technological structure – an interim solution or an historical accident growing at a frantic pace by default because the 'real' Information Superhighway (whatever that is) has yet to arrive.

The Internet is not only here to stay, but is well into a sustained high-growth phase stimulated by an application analogous to the 'graphical interface' on the PC. This is the Worldwide Web, through which the Internet now presents its most popular face to the world.

So successful has the Web become that the IT, telecoms and broadcasting corporations which all want to direct the traffic across the Information Superhighway, are coming to believe that it now has the critical mass necessary to provide the basis for at least the first generation of interactive services.

History

What eventually became the Internet started life in the US military. Conventional 1970s data networks were host-oriented with a central point of failure. Come the nuclear holocaust and one hit could freeze all military communications. The solution involved a mesh of autonomous network elements. If one dropped out, the others would automatically reconfigure themselves and route data around the dead sector.

By the 1980s the Internet was functioning as an academic network, linking US universities. Its resilient, modular nature meant that it could grow in a relatively random way, with all the bits owned by different organisations. Then it just kept growing.

The Web: what is it?

The Worldwide Web is essentially a way of making documents available on 'Web servers' under a simple, common standard on computers all across the Internet. Gone are the arcane commands necessary to navigate between resources. The Web is set up in a way to allow users to 'browse' with a graphically-based interface providing a familiar 'point and click' environment.

Invented by a British researcher, Tim Berners-Lee, the Web is based around the concept of generalised mark-up, already well-established as SGML (Standard Generalised Markup Language).

SGML text files are embedded with tags which define the content in a 'generic' way – independently of the 'procedural' instructions required, for instance, within a word-processing document to specify type size, design or exact position. SGML documents are then able to be stored and accessed in a structure defined by their content, and are portable between platforms and applications.

The 'Web's Hyper Text Markup Language (HTML) shares the basic principles of SGML. Text within documents may be tagged in a hierarchy of headings, and other tags may be used to give emphasis within the text body.

But the central innovation of the Web is the technique which allows one document to be linked to another. Words or phrases within a document act as links to other documents.

There is no hierarchy to provide a context for material; instead the participants themselves develop documents containing lists of other documents. In other words the navigation structure is designed to evolve organically as needs and imagination allow. The embedded tags define a Unified Resource Locator (URL), which is essentially an address locating a single document somewhere on the Web. To the user a link word or phrase appears as a highlighted string on the screen.

Design

Sound, pictures, diagrams or even animations may be supported within Web documents, but because these tend to slow down the application, designers are always treading a fine line between the temptation to enhance the pages (and therefore add considerably to the file sizes) and the need to keep the pages lean and increase the chances of people logging on (and coming back).

The villains of the piece are the graphic files, which are usually converted into a highly-compressed format, but can easily be many times larger than the main document. These remain separate files associated with the HTML files on the Webserver (or even another Webserver), and are referenced and positioned by HTML commands in the document.

Much ingenuity is often used to select or doctor graphics so that they become more 'compression-friendly'. Page designers might spend hours removing extraneous detail or colour gradations so that they may shave a few kilobytes from the file.

Users

From the potential user's perspective, the Web is accessible through a 'browser' which may be downloaded free from the Internet (although this is beginning to change as the field becomes commercialised). Browsers are available for most workstation/PC platforms.

The browsers interpret the content contained within the HTML files into a procedure using the specific workstation resources. So the appearance of documents can differ between terminals, it being up to the user's system to specify fonts, background colours, some sizes and so on.

Differing screen sizes are another problem – as HTML deals in a mixture of absolute and descriptive data, care must be taken in mixing graphics (delivered in an absolute size) and text (which flows in according to the margins set by the browser). New HTML features allow designers more completely to control the appearance of their pages on the destination screen.

Designers who want to control the appearance of the pages by trying to associate text with specific graphics quickly find it easier to embed the text in the graphic itself. It is no longer hypertext, but it is possible to specify this graphical text as a series of 'hotspots' which may be 'clicked' by the user to move on to other material or trigger some sort of interactive procedure.

Useful animation is available but slow, while real-time voice across the Internet is already a reality. At least two companies are offering voice communications between microphone-equipped terminals. Although AT&T is not expected to go into financial crisis immediately, the idea of international calls at virtually no cost are causing ripples of concern in the telecoms world.

Any assessment of the Internet must recognise that today's scene is a snapshot of an evolving process. The Web has both a critical mass of users and a growing army of enthusiasts to develop the applications. It should be thought of not so much as a network or even a technology, but as a huge user-base locked together in a development process.

Some acronyms

The basis of the Internet is TCP/IP (Transmission Control Protocol/Internet Protocol) and Unix (originally an operating system for minicomputers). The Internet Protocol basically formed the Unix machines into a network with a global naming scheme and allowed packets of data to flow between them. The TCP portion is a higher level protocol which orchestrates the flow of data in and out of each computer.

The key to IP is the 'router'. Routers are network elements which act like a series of intelligent traffic lights, each set responsible for its own intersection. Each router in the Internet maintains a routing table which tells it in which direction it should send all the data packets it receives. It maintains the table by chattering constantly with all the other routers around it. If congestion should occur on a specific route, or if another router should go down, it will automatically update its table.

Other useful acronyms include:

PPP: (Point-to-Point Protocol). Can be used as an access protocol for dial-up access to the Internet.

ISP: (Internet Service Provider). Offers access on commercial terms to the Internet.

Content

As the tabloids have been quick to sniff out, soft and hard porn are definitely a feature of the Web, but then porn magazines and business journals sit side-by-side in the newsagents. The Web is a medium, not a culture, and it is now moving firmly into the mainstream.

For instance, IBM, Microsoft and Sun are all moving heavily into the sector: IBM has included browser software as standard in its OS/2 operating system; Microsoft has practically given up its strategy of building a parallel global network which will interconnect with the Internet, but has decided to compete directly with Netscape in the browser market, and Sun has invested heavily in Web Server software and authoring tools.

Security

Web sceptics point to the Internet's lack of security and its unreliability. Systems and procedures are available to insulate corporate networks from unauthorised access via the Internet.

These include installing a variety of systems which monitor traffic, right through to what are called proxy server systems. These expel resources available from the Internet (the corporate Web pages, for instance) to the outer reaches of the corporate network and make users connecting to the Web from the corporate network use IP addresses located on these servers. This makes it harder for so called hackers to penetrate the corporate network from the outside.

The security of data travelling across the net, particularly credit card authorisations, can be a problem (but then, so can conventional methods of using credit cards).

Information provision

Until the advent of the Web, providing on-line information involved proprietary systems, intermediate information providers and on-line bureaux. Unless you were really 'serious', with some valuable information to sell and a way of marketing it, going on-line was fraught with difficulty and expense.

Becoming a Web information provider on the other hand, is a relatively trivial affair. Again, there is no necessity to dedicate Unix facilities: most platforms, including PCs and even the Macintosh, are supported, so that for minimal outlay any company with a little in-house expertise can create a Web presence.

The single biggest cost is providing leased line access. LAN-based email systems for instance, can be seamlessly hooked to the Internet via the services of an Internet Service Provider (ISP). The ISPs will provide the domain names and arrange the connections. They will also provide for ISDN or dial-up links. Most ISPs also provide space on their own servers for customers wishing to have a presence.

Some companies are using Web software to provide an internal information system. This also provides a handy proving-ground for those unready to open their doors to outside access.

New Phase

There are different ways of measuring the success of the Internet/Worldwide Web model.

The fact that the Internet, the network, is itself very successful, is just one.

Perhaps it should really be measured as a set of technologies and skills: in IT nothing succeeds like success. The dynamics of a complex market, where it takes the efforts of a whole range of specialist companies just to get one product to market, are further accentuated when that product has to be compatible with a single network serving millions of terminals.

Like iron filings obeying a magnet, the relatively sudden success of the Web has seen IT companies arrange themselves into an intricate support pattern to give a broad range of existing IT sectors an Internet dimension. Fileservers are optimised into web servers, modems are specially packaged and often sold through Internet Service Providers, network security specialists have vied to create a multiplicity of Internet 'security' products for corporate networks. But while the original focus of the corporate network vendors was grudgingly to provide an Internet prophylactic for the corporate network while still allowing intercourse between the two domains, a new approach sees the Web being brought right into the corporation.

Intranets

Many companies are developing what are now called 'Intranets'. This is where a Webserver is set up on the corporate side of the private/public network boundary to provide new services for corporate network users.

Increasingly sophisticated Webserver software provides an ideal basis for a whole range of enhanced intracompany applications – things like company product information, telephone listings, past memos and material on company procedures and so on, can all be Webbed and, better yet, kept constantly up-to-date. The communication can also be two-way by using the Web's form features to gather information back from staff.

Where users already have familiarity with their browsers and web conventions, the costs and disruption involved in introducing what used to be called the 'paperless office' can be minimised.

But perhaps the key driver on the Intranet application is the open and non-proprietary nature of the system and its software.

Corporate Intranet developments are often driven by IT staff who have acquired their skills from their own interest in the Internet. The low cost of providing browsers and building the system (often using tools of various kinds which can be downloaded free from the Internet itself) removes other barriers to Intranet introduction. Most of all, there is a confidence in the future of the approach. Not only is it cost-effective now (in comparison with proprietary electronic office solutions), but a collective expectation that the broader Web will continue to evolve means it has a limitless growth path into the future.

The rapid growth of Intranets can be interpreted as the ultimate endorsement of the open network system approach. Intranet products are now being heavily promoted by all the heavy-weight computer and networking vendors, including Digital, Hewlett Packard and IBM. If this trend continues IP and all its associated protocols and applications will become entrenched standards in both public and corporate networking domains. As a focus for cable service development the Web is now impossible to ignore.

9

Fog and the Shapes within

T here is very little mileage in trying to predict the specifics of the technological future – such speculation is almost inevitably wrong – at least in detail. And detail, as we know, is where the devil lurks for companies scheming their strategies.

The history of technological prediction gives little comfort.

There has been the occasional spectacular success. Nearly everyone knows about Arthur C. Clarke's brilliant prefiguration of the communications satellite. But Clarke's feat can be seen as the exception that proves the rule – a 'one-shot', as it were, made all the more remarkable by his inventing the concept at the same time.

The problems of prediction

Vague destinations can be discerned. It is in trying to chart the geographical detail where futurologists and research companies usually come undone.

Having grasped the concept of the helicopter in the 1930s, futurologists plotted a linear progression for its application along a line then recently driven by the car. In the 1930s, of course, the car 'concept' was still only 30 years old, and most of the cars that had ever been built were probably still on the road. Within easy living memory, then, the automobile had advanced from being a contraption required by law to travel at low speed preceded by a man with a red flag, to a relatively accessible consumer item. How much more likely it must have seemed then, than now, that the recently invented helicopter would follow the same track and soon be parked in every garage.

How extraordinary it is that the car has remained, different in every particular, but virtually unchanged in its appearance, function and 'user interface' a full 60 to 70 years on, while the helicopter is one of the least accessible forms of transport available.

In the chaotic world of networked multimedia, the accelerated rate of change also makes nonsense of predictions.

Take video-on-demand and multimedia business convergence.

The catalyst for what developed into an avalanche of speculation was a proposed merger in late 1993 between two US communications giants: TeleCommunications Inc, the biggest US cable operator, and Bell Atlantic, one of the regional telecoms operators created at the break-up of AT&T in the early 1980s.

With an important campaign strand of the 1992 US presidential election built around vice-president Al Gore's promise of a government inspired 'Information Superhighway', the announcement of a mega-merger – at the time the largest ever attempted – provided what we in the media trade called a 'peg' to build a whole range of broader stories about the new technologies which seemed to be pushing formerly discreet industries like broadcasting and telecoms closer together.

With broadcasting going digital and telecommunications theoretically able to provide high-bandwidth digital services into the home or business, both sectors suddenly looked very similar in terms of technical capability.

The sheer size of the TCI/Bell Atlantic merger therefore leant credence to the idea that a major shift in the market was about to take place, with new, multi-media communications companies being formed to combine telephony and entertainment. Why else, the reasoning went, would these corporate giants – with all their research resources and know-how – be willing to risk billions of dollars?

Why indeed?

The merged giant's erstwhile competitors moved quickly – other mergers were mooted and extensive interactive television trials were announced.

Then the TCI/Bell Atlantic merger was called off, just months after it was announced. This took the frenzy out of the media coverage, but despite this the raft of technology trials already unleashed continued. Here in the UK, BT was especially smitten by the bug – it announced a trial of video-on-demand (VOD) with participating employees.

This trial went into a second phase to test applications and BT says it will look carefully at the results before making any decisions about whether, how and where it is going to deploy service. But in the US, many trials launched in the wake of the merger that wasn't, were subsequently quietly scaled back or ended.

Chaos

Clearly, advances in information technology do not deterministically push the resultant applications down a line of least resistance. Many chaotic forces are at work.

Like astrologers, futurologists are on firmer ground in predicting a general shape – the art of making the obvious seem profound. But even so, like astrology, endless predictive failure seems unable to undermine the demand for more reports, growth projections for specific technologies and knock-on scenarios that fail to materialise.

But perhaps such analysis misses the point. It is often the most rational, 'thinking' people who require some sort of roadmap into the future. By its very existence and the faith being placed in its outcome, the 'plan' becomes a self-fulfilling prophecy.

Something of the kind is happening with the development of the Information Society, as it is called in Europe, the Information Superhighway, or Global Information Infrastructure as it is variously known in the US.

These ideas might be better viewed as aspiration rather than prediction.

At a political level they provide a dynamic for the national adventure – a 'big picture' to which a specific political framework or term in office can be hitched. In the UK, the best example was Harold Wilson's first labour government in 1964 which attempted to invoke the 'white heat of the technological future'. Labour would embrace the new, would propel the UK into a new phase of growth by understanding and promoting policies which could capitalise on technological change.

It didn't happen, but at least it put a positive destination on the map.

Epochal expectations

Both political and industrial initiatives are underpinned by epochal expectations – history constructed in advance. Our current collective expectation is for an 'Information Age' as an inevitable consequence of the so-called information revolution we are currently experiencing. This construction uses the only doubtfully valid historical concept of the preceding agrarian and industrial 'ages' as a template on which to impose the next.

It is received wisdom to extrapolate a whole range of specific impacts flowing on from what is now viewed as a solid fact: that a huge 'information' industry will transform the nature of work for most people, at least in the developed world.

The predicted imminent development of an 'Information Society' is leading educationalists to recommend very specific changes in approach for people currently in secondary education. That means that radical changes in the job market are expected to develop over 10 to 20 years. But it is the specificity of the vision that should give pause for thought.

For instance?

Does it really follow that networked information systems will promote the currently perceived tendency toward flexible, contract employment? How much is this a result of changes in political climate and specific employment laws, rather than any workplace re-engineering flowing on from technological changes?

It is just as possible to argue that, far from technology freeing us all to work as semi-autonomous individuals, the sheer social and occupational promiscuity brought about by us all in effect, sharing the same office across a vast global network, might lead the powerful organisations actually to impose more, rather than less, security of employment. To provide less would be to invite chaos into the crucial areas of intellectual property, company sensitive information and so on.

If work becomes an electronically tradable commodity it is reasonable to assume that moving employment will be easy – but it may not lead to a nirvana of 'casualised' intellectual work.

According to one popular version of life in the Information Society, our expectations of lifetime employment will be distant history. Instead, all employees will also be contracted to at least two other organisations. In a society of uncertain employment tenure, electronic workers tend to hedge their bets so that if one job folds, two others remain to keep the revenue flowing while a third is found.

But will this really be so? Let's take an example of how such a situation might work in practice.

Company 'a' is a huge global advertising agency with thousands of employees located around the world – let's call it Starsky and Starsky.

Starsky and Starsky has grown rapidly through the preceding decade, but growth is no longer meeting shareholders' expectations and there is the small matter of a gigantic debt burden, incurred through a series of acquisitions.

Major shareholders force a boardroom coup – both founding Starskys are asked to take a large amount of money and go quietly. They don't.

Instead, they simply set up shop in a different, but equally accessible part of cyberspace, and re-hire thousands of those loosely contracted employees to work for them.

Given this sort of threat (you could imagine an infinity of similar scenarios) to the network-based 'virtual' organisation, it is at least reasonable to argue that because the technology allows participants to move more easily from employer to employer, or between client and client, there may develop a countervailing tendency for employers to demand tighter, rather than looser, employment arrangements.

Just as globalised electronically-based currency markets can have a tendency to accentuate market movements, and therefore lead to pressure for countervailing regulations or new financial 'instruments' which mitigate their effects, so electronically-based work may in practice cause personnel instability in certain circumstances causing employers to stabilise their contractual arrangements.

This is not prophecy, by the way. It is simply an alternative scenario. If my projection eventually proves closer to reality than its alternatives, then it was simply a lucky stab.

Intellectual capital

There is a pronounced trend in what economists call capital formation occurring around the ownership of intellectual property. IT of various kinds is allowing companies to identify, trap and then distribute the knowledge and expertise that they had always had but which, until the advent of high technology, they were unable to quantify and package in such a way that it could be transferred or sold. It either moved on with individuals to new companies in the form of 'experience', or remained in an organisation's collective subconscious, only to be teased out at meetings or in written proposals and reports – one-off airings as it were, and probably not all that consistently presented.

For both technological and cultural reasons, organisations have simply got better at collating intellectual property and packaging it. And where there are sellers there are usually buyers – it is no accident that consultancy is now one of the world's fastest growing sectors. In many ways it is closely related to IT itself, and is dominated by increasingly global organisations dedicated simply to collecting knowledge and selling it.

Knowledge management experts maintain that we're still at the tip of the iceberg. On average only a tiny percentage of so-called 'intellectual capital' inside businesses is currently being tapped. They suggest that the future growth of the information society will involve harvesting this hidden resource in a myriad of ways.

Viewed in this way, the individual expertise and experience you and I gather as a sort of by-product of our working lives as we climb the slippery corporate pole, is fast being viewed as a corporate resource. At present it is protected, where at all, only through inefficient traditional contract arrangements involving confidentiality. But increasingly, employment contracts are likely to include set time-period prohibitions on working for competitors.

Viewed from the centre of this trend, that future working life of loose, technology-enabled work patterns looks a little less inevitable.

Too many variables

The simple point is that there are a host of variables to be grappled with before you can even begin to project the shape the Information Society. Idle speculation is fairly harmless, but advice to school children on their educational needs, with all those variables in the way, should be treated to closer scrutiny.

In the case of the 1930s helicopter, it is now possible to speculate that safety and environmental considerations made the low-cost family helicopter unable to get off the ground. But in the 1930s and 1940s the idea was not really so ludicrous as it now seems. After all, who can doubt that automobile-based road transport, if mooted for the first time in 1996 as a feasible system, would be laughed off the stage. Self-steering metal boxes travelling side-by-side at 70 miles per hour? An estimated death toll in the UK of something like 5,000 per year with many more serious injuries? I think not.

There are examples also in the more recent past. By the mid-1980s conventional wisdom had it that increasing computerisation would create a huge jobs market for programmers. In a strikingly similar way to the current boom in information working, curves were plotted, assumptions were made and the conclusion was that literally hundreds of thousands of programmers were going to be required in the immediate future. School pupils and parents were told that computer literacy and programming skills were going to be essential for employment.

Nobody was ever harmed by learning a programming language, but the computer industry moved on to high level languages, the PC and packaged applications. Those countries who took the concept the most seriously, like Ireland, ended up with a surplus of programmers.

The real knowables in the area reduce to a simple truism – that the rate of change continues to accelerate; that it is driven, at present, by technological developments and that it tends to ripple out in completely unexpected and unpredictable ways.

A general direction is discernible, but developing a framework which attempts to predict the specifics must be viewed as a hazardous business.

So what are the general shapes?

What can be predicted with some degree of certainty are 'building blocks'. We can't tell if the helicopter will become a consumer item and cost £14,000. But we can understand the general direction of things by getting a feel for the underlying technical improvements – these have been the one set of constants in an otherwise chaotic world.

If we use the Internet as an indicator, rather than a model, it might be possible to identify a few broad (very broad) dynamics.

Distance nobbled

The old communications charging model which assumed that the further the signals has to travel, the more the customer has to pay, has been severely challenged by the Internet. In fact the Internet has not 'discovered' distance independence. The underlying economics of the global telecoms network have simply been bottling it up for a couple of decades because it suited everybody except the customer to do so. But the Internet has illustrated that simply charging for access at a flat rate works as a business proposition.

Now that the genie is out of the bottle, the expectation is that charging for capacity will disappear almost completely. According to the World Bank, for instance, capacity is expected to cost 'near zero' by 2005.

Instead, charges will chase 'real costs'. As the cost of raw bandwidth is diminishing as a proportion of the total real costs of running a communications network, pricing will be forced by remorseless competition increasingly to peg itself to other measures of usage. These may include enhanced services, customer premises equipment rental (as this rises as a proportion of costs) and network access (line rental).

The trend is clear but the timing is still worth some argument. The real market cost of bandwidth may well fall to 'near zero' as a commodity traded between telecoms companies, but it is difficult to believe that 'near free' bandwidth will uniformly trickle down to all customers by 2005. Most countries still have state owned and operated telcos after all, and many of these are still struggling to provide a phone penetration level of 5%.

But distance barriers are being flattened across the network.

Chaos

The Internet is entering its most exciting phase as it attempts to build some sort of structure around which services might be delivered.

In this context the 'debate' has shifted significantly over the past year. Much to the chagrin of the cyber romantics, who previously rejoiced in the notion of a flat structure, the Internet is beginning to see companies attempt to gain footholds in what they expect will be a lucrative position in an emerging value chain.

Just as an indicator to how quickly real events make a mockery of prediction, corporate strategies were very recently predicated against the idea of de-intermediarisation. The Internet would be able to strip away the complex distribution chains found, by necessity, in the physical world. Individuals would be free to communicate directly. Goods could be ordered direct from the factory, information and art could be disseminated on a democratic, one-to-one basis.

The theory only ignored the fact that most people have no desire to communicate directly – at least not for everything. Intermediaries are not leeches in the distribution chain: they perform all sorts of necessary functions on behalf of both producers and consumers.

Far from removing intermediarisation, the Internet looks likely to intervene at a myriad of new value points.

10

Cyberspace for sale

The Internet represents a major opportunity for cable operators. In the longer term, the applications running across this global infrastructure will allow them to cost/justify the roll-out of high-bandwidth services to the home and business. In the meantime they have the opportunity to forge important relationships with content providers keen to take advantage of cable's early entrée into broadband delivery via the cable modems which some UK cable operators may begin deploying in 1997 (see Chapter 11).

But there are also immediate opportunities in dial-up Internet access. By careful tariffing and promotion cable operators can use dial-up as a way to strengthen the overall cable 'package' and build experience and a customer base.

Possible approaches could include special tariffs for access using cable lines, optimised to increase usage at particular times of the day.

At this stage in the Internet/cable relationship both sides are wary of the implications of flat-rates. From the Internet Service Provider's (ISP) perspective, free calling could encourage excessive connect times which could tie up expensive modem ports at the access nodes and diminish the provider's ability to provide acceptable availability to the subscriber base. In practice, users soon discover the attractions of keeping the dial line as a permanent circuit to the ISP; the ISP has to invoke counter measures, like disconnecting after an acceptable time period; users learn to connect periodically and disconnect; and so the battle goes on toward the inevitable falling out between provider and subscriber. Cable operators are also wary of the same effect on the switching facilities.

Cautious approach

So far, the cable response has been mixed. Telewest has explicitly excluded ISP numbers from its free-calling tariff, while Videotron provides free calling between its own ISP and cable telephony subscribers.

Cambridge Cable experimented with the concept for a time by allowing a local ISP to include dial-up users in its Centrex group, thus providing free-calling.

However, at the time of writing, the attitude is still equivocal.

One approach is to build a separate company. Telewest's Cable Internet subsidiary has signed a deal with the Internet Service Providers Consortium (ISPC). This organisation was formed to represent the combined purchasing power of around 20 small Internet Service Providers (ISPs) who were given notification of the termination of their access arrangements with PSINet, an international Internet Service Provider, after EUnet GB, the original provider, was taken over by PSI.

Under the arrangement, Cable Internet will provide the Internet connection points at an agreed rate to consortium members, nearly all of whom have elected to take the service.

Taken together, the ISPC connects over 22,000 individual and corporate Internet users and claims it is probably about the fifth-largest ISP force in the UK, after PIPEX and Demon, PSINet and BT. Most ISPC members have a single dial-up site or Point-of-Presence (POP) and have built businesses connecting users within their own local dialling areas.

Cable Internet says that it will work at putting together cross-selling arrangements between Telewest and other cable company sales teams, and ISPC members.

Cable on-line

Cable On-line was launched in November 1995 and is a sister company to Cabletel. It is aiming for nothing less than being the UK's number one Internet provider within two years. It will shortly have 70 points POPs across the UK and it aims to be an important wholesaler of Internet services to other UK cable companies which it expects will offer Internet access under their own brands.

Cable On-line is available to anyone in the UK (not just cable subscribers) but it is likely to be marketed by participating cable companies as part of their services portfolios.

Cable On-line has identified service quality and support as the keys to gaining share in this emerging market. The basic service is priced at £12.95

per month with a one-off connection fee of £20.00. Service features include an 0800 number for free technical support (90% of calls being answered within 15 seconds), high bandwidth connections to the US (where the bottleneck occurs at the moment) and 99% availability of both host modems and Cable On-line servers.

ISDN and leased line access will also be supported and 90% of the UK population will be able to dial a POP with a local call.

As the ISP market matures players are tending to focus on service quality to keep what have proved to be notoriously fickle subscribers. But at the same time, there is a limit to how far price competition on the flat-rate access charge, typically £10 per month and below, can go – ultimately, as the dial-up Internet user base increases, the competitive pressures on all sides must see some major service packaging.

The customer

An Australian on-line survey carried out through May and June of 1996 found that most Internet users go on-line once a day and expect to increase the amount of time they spend on the Internet over the next year, in part because they're ready to make use of the wider range of services they expect to find there. At present entertainment, email and research are given as the main activity, but a large majority of respondents said they would be willing to try on-line banking and shopping.

Clearly a major part of the attraction of the Internet is its low cost. While users clearly value it highly, that doesn't mean they are willing to pay more to get a better service – far from it.

While the network's response times are always given as a major concern, cost is apparently an even greater concern. Very few respondents said they wanted to pay more for a faster service.

Such attitudes provide a clue as to the sorts of results the interactive TV trials are currently getting. Information and entertainment arriving on a screen is not yet perceived as something that should be paid for, at least by the user, above a certain trivial level.

In fact interactive TV trial results may be even more downbeat than the attitudes of Internet users, given that these are a very skewed population sample – in the Australian survey 87% of the respondents were male, most were aged between 20 and 44, a majority had tertiary qualifications, and their incomes were well above average. This picture accords with Internet users globally.

It is possible that attitudes to cost will change as the Internet offers

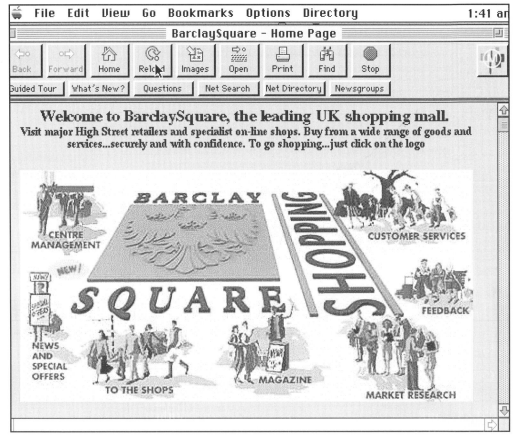

BarclaySquare provides an early version of a 'virtual' UK High Streeet.

electronic equivalents of what are perceived as high value real world services like banking and shopping. But even this shouldn't be assumed.

Small change

Shopping across the Internet is traditionally held out as a potential killer application. At present it tends to be clumsy in operation and unspectacular in its commercial result, although its supporters point out that the availability of electronic cash will transform its prospects. Perhaps.

According to Verdict, a UK-based retail consultancy, on-line shopping services are "unfriendly, cumbersome, painfully slow and inconvenient", a reality reflected in consumers staying away in droves. According to Verdict, the Argos on-line catalogue, available at an Internet-based virtual shopping mall called Barclaysquare, managed to sell just 22 items in nine months of operation.

The consultancy asserts that electronic shopping will have no impact in the

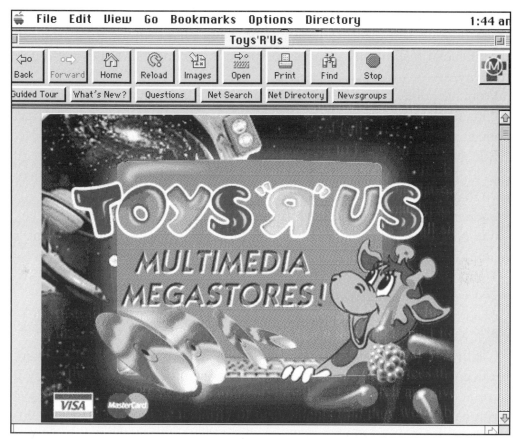

Despite the hype, some analysts project a low demand for on-line shopping.

next five years and a minimal impact in the next 10 years. Electronic shopping's £42 million in UK sales in 1995, projected to rise to just £55 million this year, looks set to remain a tiny fraction of the £6 billion home shopping market (catalogues etc) and an even more minuscule proportion of the total annual UK retail spend of £160 billion.

Not so fast..

But not too much credence should be placed on linear projections running into the next century – much can happen in such a time span.

The Internet has the ability to upset traditional value chains within a sector – especially where that sector has seen the development of specialised third party data processing or network services.

The new power to weight ratio of credit-card authorisation companies is an oft-quoted example.

The banks and credit card companies were originally happy to cede responsibility for credit checking – the function seemed most appropriate to an independent clearing house, processing credit information from all the financial institutions without competitive commercial issues getting in the way.

From the point of view of the banks, before the Internet, the credit checking companies were nicely 'contained' in a *cul-de-sac*, well insulated from direct contact with the customers.

But suddenly the popularity of the Internet may mean that the credit checking companies have both the means and the know-how to offer their own loans and services direct to their customers' customers.

Suddenly, that boring old credit information business now occupies a key value position in the new matrix. After all, if you want to build a business lending money you need:

1/ money;

2/ a way of reaching customers,

3/ a low cost, and/or high service level proposition to attract them, and

4/ a way of assessing the risk of individual loans.

Credit checking companies already have **4** and are skilled data network service providers. If the Internet can provide them with **2** and **3** then they'll have no difficulty raising enough **1** to develop the business.

But naturally, this power pendulum can swing both ways. Because credit-worthiness data is so obviously key to banking and a whole range of other retail activities, the real originators of the credit data (ie the credit checking companies' customers) might think carefully about its value and under what conditions they pass it on to third parties, if only because they too can use the Internet to set up new, multi-lateral arrangements to share their own bad debt information.

This way they can win three times over. They use the Internet to share bad debt files and not only reduce their own credit checking costs but begin selling credit-checking to third parties. At the same time they prevent their own data being used against them.

Cowboys

There are potential versions of this sort of matrix formation in just about every sector: insurance, securities trading, consumer retailing, media, entertainment and so on. As the example above shows though, it is often difficult to assess the ultimate balance of advantage to each player in the matrix.

In such circumstances, parties can be nervous about making an early move in case all the other players combine against them.

This scenario will be strangely familiar to comedy western fans. Cowboy One pulls a gun on Cowboy Two who puts his hands up. "Not so fast.." says a voice from behind as Cowboy Three enters the room, gun pointed at Cowboy One. "Not so fast," says Cowboy Four aiming at Cowboy Three from the open window.

Do they all start shooting? Or do they all slowly put their guns down and come to an arrangement?

The new paradigm

Until the Web, the on-line industry pursued the idea of the Technology Solution. Facilities had to be presented as fully-formed, complete, services – or not presented at all. The reasons for this approach are not hard to find. On-line services naturally started life with high value commercial applications, like electronic funds transfer and electronic document interchange (EDI). A culture developed which valued security and resilience as top priorities in both overall and specific design.

To develop tight, function-specific systems, like those designed to move large amounts of money around the globe, meant designing out any flexibility in the way the system could be used – you did not want someone designing a way to run off with the money.

The Internet goes to the other extreme by allowing the 'systems' to be built on top of the network, rather than inside it. This has allowed participants to develop exceptionally complicated functional relationships between chunks of data sitting on different parts of the Net.

The fascinating part of this dynamic is not the sophistication of the technology, but the way this aspect has accelerated the evolution of services.

On-line laboratory

The net is like a supercharged laboratory which deploys new technologies and services; tests consumer attitudes towards them; and then refines the approach or sees it sidelined by a new approach, all in a matter of months or even weeks. An extra frisson is added to the marketeers' burden by the ultra-critical reception new services can get. For in this stage in its development the Internet still retains its Utopian strain and many Net users are alive to any transparent privacy transgressions or instances of commercial gerrymandering.

An uneasy balance is constantly being refined between commercial and private interest. In this context, much power still remains with private users who are able to combine, using the Net itself, to boycott or undermine services which are deemed not to come up to basic ethical standards.

Advertising is now well developed (not so long ago, a very contentious issue), but as in the paper world, the gerrymandering of what purport to be commercially 'disinterested' services is often singled out for special attention.

One major ethical fault line has showed up with search engines.

Unacceptable behaviour

The search engine is now just one of many service types which attempt to mediate between individual Web users and the near infinity of Web-based material available for him or her to 'browse'.

The search engine maintains a huge index of Internet accessible material. The idea is that users log on and perform a keyword search – material or sites claiming to possess material conforming to the search terms is flagged up and the user can surf off to inspect them

But as the sheer number of sites and the accompanying bulk of material grows, even the most refined set of search terms is likely to throw up a huge list of possible sites to visit.

Inevitably, commercial temptation has reared its ugly head. At least one company has begun overtly to sell guaranteed positions early in the list should the site or its material be selected by a search term.

This is the Internet version of display ads in the Yellow Pages, or calling your company AAA Aardvark Services. As a result outraged users quickly formed a boycott and organised an email campaign to fight the practice.

At least paying to be 'Aardvarked' echoes behaviour in the paper world and its implications are well understood. But the Web seems to be creating new categories of service that have no close equivalent in the analogue world. On-line banks and shopping malls are self explanatory, but what is a 'Profiling Service' or a 'Personalised Service'? And how do the ethical issues impact on them?

These services have evolved to tackle two functions. They provide a way of furnishing Web sites with visitor demographics without the Net surfer having to fill out a laborious form every time a new site is visited. By registering your demographics once, you get to visit participating sites without encumbrance – you either fill out a simple password, or access is granted from the intervening Web site or even from data embedded in the browser. The user's demographics are registered automatically.

Now there are serious attempts to take the idea a step further by refining the profile to the extent that it not only allows you onto a site, but actively

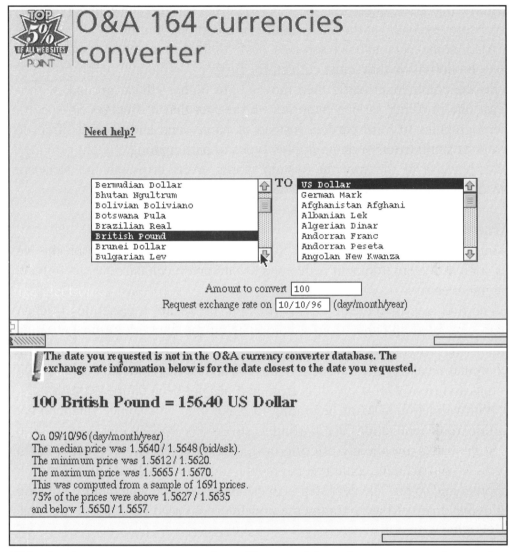

This simple currency converter from Olsen and Associates may not change the world, but it is an example of a perfect Web application.

presents which sites or which specific material is likely to be of interest.

This is an application of the "intelligent agent" function. In some systems the Personalised Service maintains your bookmarks as a personalised collection of data on its own server and then adds other bookmarks which it works out that you may be interested in.

Like the search engine service, the Personalised Service opens out other commercial possibilities for the provider – it offers another 'power point' in the rapidly-evolving structure building amongst content and service providers. Just as sites offer banner space and search engines use commercial criteria to position entries, so these personalised services have an opportunity to ensure 'eyeballs' at the site's launch or on an ongoing basis.

Also, some personalised services have used the valuable demographics that, by definition they must collect, for further commercial purposes.

So the commercial battle then moves onto higher ethical ground, with at least one profiling service majoring on the fact that it divulges no specific demographics to third parties. Instead of names and addresses it offers its users as a collection of demographic types to participating content providers and, because of the way the system works, even strips out the Netscape header when the users go on to access the site.

Transactions: the missing link

Internet enthusiasts are convinced that commerce will take off on the Net as soon as the environment is deemed secure and mechanisms exist to make payment easy.

The first step is to create an environment for secure transactions – where data generated and sent from a computer on the network could get to the other end without being intercepted. From the opposite perspective the computer receiving the transaction must be able to authenticate that it is valid – that it is from who it says it is from and has not been tampered with.

With validation systems beginning to make their appearance financial and commercial applications are starting to stir.

Some banks are already offering on-line banking services, where you can browse your account, authorise payments, move money from one account to another and so on. The next step is to establish systems which close the loop, allowing Internet-based commerce and Internet-based banking to stimulate each other's growth.

Currently, Net payment for goods or services usually entails credit card authorisations which can be inhibiting for the payer, especially if there is no real way of knowing where the authorisation is going. It could be argued that credit card authorisations are just as likely to be abused back in the physical world, but somehow launching your number into cyberspace seems more dangerous.

The gleam in the entrepreneur's eye is the idea of establishing what would effectively be a borderless currency.

Progress seems to be relatively steady, rather than explosive, in this key area. In late 1995 the aptly-named Mark Twain Bank, based in St. Louis, Missouri, branched out on the electronic frontier by becoming the first bank to offer an electronic cash facility. Users are able to maintain an account on-

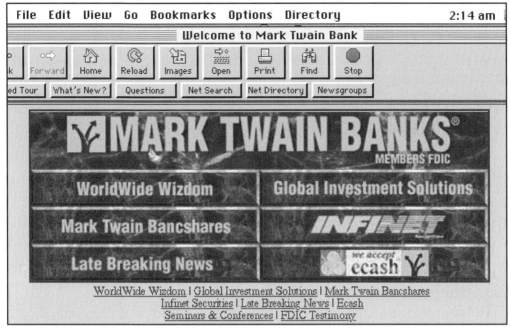

The electronic 'lobby' to the Mark Twain, just one of many banks beginning to explore global retail banking.

line and pay for goods and services available on the Internet using a Dutch system called Digicash. This is a debit system, an on-line version of smartcard-based electronic cash and will probably, if all goes exactly to plan, merge with it through on-line card readers/encoders. A year later the bank was still sounding bullish, but was anxious to explain that the service was still at an early stage.

The Mark Twain Bank says that people are more comfortable with relatively tangible electronic cash, which travels about anonymously, as it were, just as notes do. The electronic coinage is exchanged for 'real' cash at a 'real bank' and travels about across the network until a user redeems it for real cash again, at a participating bank. Users apparently like electronic cash better than credit cards because they feel less exposed, can buy anonymously and can't overspend. Digicash can also be exchanged between currencies.

The prize, again if all goes according to plan, could be quite staggering. If the sector begins to grow solidly, a virtuous circle could begin ticking over – increasing numbers of people use on-line banking, then gradually begin using Digicash, or its competitors or replacements and begin buying the odd good or service across the Internet.

Then, as the number of users grow, it becomes a low-cost payment option for all sorts of regular or irregular private commercial transactions –

telephone and electricity bills, hire purchase and so on.

The questions facing banking regulators are profound. Electronic money, by definition, becomes another currency and has an exchange rate with all the other currencies. As described it looks like a wonderful way to launder money. It might be an ideal place to 'hedge' against currency movements. There is no doubt an army of international financial regulators quietly watching and listening to developments.

Security

This has to be kept in perspective. There is obviously potential for fraud, but such an observation is only meaningful if it can be compared against the risk of the alternatives. Apple's Director of Cybertech Products, Doug McLean, speaking at the Telecom 95 Internet conference, claimed that there is more conventional credit card fraud happening in any one day than has occurred on the Internet during its entire lifespan.

He says that research indicates that Internet fraud will represent £1 per £1000 transacted, while standard losses are currently estimated at £1.41 per thousand. You pay your money and take your risk.

Advertising

Advertising hasn't really worked out a role for itself. All agree that it will play an important part, nobody is quite sure how and when. At present the most obvious mechanism is to sell 'banners' on pages which receive a lot of 'hits'. So ads are often situated on welcome pages, search engine pages or menu pages. The idea is that the browser clicks on the banner and gets a screen-load of further information from the advertiser's site.

As in print advertising, the trick is to associate an ad with content likely to attract the target customer and to expect to pay more for positions which attract the most 'eyeballs'.

Voice over the Internet?

Much of the discussion on Internet voice services has understandably been about its effects on telecoms companies' lucrative international call revenues. What starts out as a basic ability for one Internet user to talk in real time to another via their PCs could have some interesting knock-on effects as the facility is exploited in new and unexpected ways.

There is already a thriving international telecoms resale sector, where independent operators provide cut price international calls. At present they do this by leasing their own circuits or by buying capacity at wholesale prices and selling the call minutes at a slimmer margin than do established operators.

The Internet may provide another more flexible and cheaper way of driving this business.

Currently there is a major effort under way to implement a protocol for the Internet called RSVP.

RSVP allows prioritisation of traffic and also provides detailed statistics to allow providers to buy and sell specific capacity amongst themselves.

At present the Internet is effectively a 'capacity limited' system. Although there is some rudimentary prioritisation, all the data wanting to travel across a particular set of links at any particular time simply has to queue up and take its chance: users modify their behaviour so that demand roughly meets supply at an acceptable performance point. If the going is slow they either come back later, switch off the graphics or just limit the number of pages they attempt to access during a session. The good old motorway metaphor applies – the greater the volume of traffic using a link, the slower all the journeys tend to be, so users tend to scale back their usage as the 'unacceptable' performance point nears.

ISPs have the ability to provide better service, on average, for their own users by buying more capacity or negotiating better reciprocal agreements with other backbone providers. But this only goes part of the way towards giving providers real control over the quality of service.

Ideally, they would like to be able to tier service quality pretty much along the airline model. Users willing to pay for speed, flexibility, reliability and comfort will pay a premium, leaving the majority to buy acceptable quality for a reasonable price. These two classes together then provide the majority of the revenue and allow economy services to be sold 'at the margin'. These would soak up the last drops of bandwidth but only provide the equivalent of an airline's 'standby' service.

RSPV allows packets of data to receive priority on their journey through the network.

With the Internet, though, one user may require all three classes of service at different times depending on the nature of the application, so providers would be able to cut the tiering cake in two dimensions.

At the end of these underlying technical enhancements will be the ability to provide relatively constant bit rates between two points on the Internet. Ultimately it all boils down to delay – if the Internet can pump data between two points and have the end systems (i.e. the computers or consumer devices at either end) encode and decode the data such that there's only a few milliseconds' delay between a voice signal going into a microphone at one end and coming out of a speaker at the other, you suddenly have a usable voice service. And you may even have a usable real-time-video service as well (although high quality video requires exponentially more data).

Cat amongst the pigeons

In the US there are already rumblings of disquiet from established telecoms operators who argue that the Internet's ability to provide voice and other services should mean that commercial Internet providers labour under the same regulatory requirements as themselves.

As an Internet user-to-Internet user application, Internet voice services probably offer little threat to established revenues. After all many would be international conversations which wouldn't otherwise take place across conventional switched links. Of course it may not stop there. The Internet could be used to provide virtual circuits for existing international resellers – users could make conventional local calls, using telephone handsets, to an access provider or reseller, and the Internet could handle the international segment of the call. These possibilities are already being discussed.

The telcos' optimistic view is that the Internet and other competitors actually help to expand the overall market. What is lost to conventional operators on the swings of directly billable call minutes is more than made up for on the roundabout of extra business from leasing them their circuits.

In any case, all expect that existing competitive pressures will cause much of the insupportably high international tariff structure to crumble. By the time the Internet is slick enough and pervasive enough to present a real threat, prices will be within striking distance of real costs in the developed world where real competition has broken out. In the mean time, all the new applications involving data networks, video and so on, will have created more demand for raw network capacity, much of which the conventional operators will supply.

The telco doomsday view is that the Internet model, with its distance-independent tariffing, will reduce the existing telcos to mere bit carriers – providing the low value, low margin plumbing for services running on top which they no longer control.

Where to now?

The question appears to be not: 'In which direction will the internet go?' but, 'In how many simultaneous directions might it go?'

The telephone giants have in the main, now recognised the inevitability of the Internet's encroachment on the business, but take a sanguine view on how life-threatening it will be. All tend to make the point that its growth allows them to sell more circuits and lines to ISPs and sell second lines and more call minutes to those using the services.

AT&T wants to take an even more proactive stance and has now begun to swing its huge research and development capability in the Internet's direction. According to John C. Petrillo, president of AT&T's Business Communications Services Business Unit, AT&T itself has a host of capabilities to bring to the party.

He and executives from other incumbent telcos are apt to point out that established telco businesses have an advanced hold over, and understanding of, their existing customers: they bill them every month, know where they live, know which ones communicate a lot, and which ones buy for value and which for price.

AT&T sees huge potential in on-line commerce. It invented the freephone concept and has seen this volume of business soar – in 1994, only $250 million in sales were made via the Internet while $100 billion was made using freephone numbers. In fact, in the US, freephone traffic had exceeded paid business telephone usage by late 1994.

AT&T sees the Internet beginning to develop as a sort of multimedia freephone service. To this end it has developed ideas which are designed to begin making the two go together.

For instance, it has a service available to content providers which would allow a Web user browsing a commercial site (say a travel agency) to enter their own telephone number in a form field and request an immediate call from a sales person.

The Web site visit then turns into a one-to-one sales encounter with the salesperson able to send images to the web user in real time.

This provides an essential 'sale closing' function on top of what is otherwise the Web's passive marketing and prospecting ability.

The weakness in this scenario is the fact that most individuals, shopping from home on the Web, have only one line. AT&T claims it has a cunning data/voice capability up its sleeve to handle this in the long term.

But if AT&T is right and the multimedia freephone approach is a potential winner, it looks on the face of it like an eventual Internet voice application, with the Internet handling both the voice and the data through the ISP (and through one connection) rather than having to require another circuit connection.

Mutiny over the bounty

The communications industry as a whole may live to regret the hype it has generated about the power and far-reaching effects of its new technologies. Interactive services will develop, but they are unlikely to have immediately profound effects on the structure of society or the nature of work – the various market shares the Internet takes at the expense of the 'Outernet' will probably build very slowly.

And the impact will be uneven – despite the Internet's growth it pays to bear in mind that less than 1% of the UK's population is currently on-line. Globally, the disparity is even more pronounced. In much of the developing world the goal is to offer the telephone to a meaningful proportion of the population, never mind access to the Internet.

But despite this, there is an expectation of rapid and fundamental change which will require elaborate regulation to ensure that all sorts of supposed social dislocation and relative disadvantage do not take place.

As the Internet is already showing, the promiscuous nature of what politicians are calling the emerging Global Information Society (GIS) or (in the US), the Global Information Infrastructure (GII) will indeed bring a slew of problems which will require legal solutions.

The politicians have decided that a concerted effort is required to forge international agreements to clarify such things as copyright, intellectual property and legal jurisdiction.

In multilateral negotiations nobody, especially not a politician, gives away something for nothing. In response to the Group of Seven (G7) industrialised nations' calls for international agreements, developing nations can be expected to make co-operation conditional on aid for their own infrastructure development.

In a European context, political interest groups can be expected to put similar social equity demands on the table. These may even include calls for universal service, or taxes to redistribute wealth from the rising information sector to the declining industrial sector.

Bit taxes

In early 1996, the European Commission funded some research into the feasibility of a 'bit tax'. The idea was to find a means of replacing tax revenues supposedly lost from taxable, conventional activity to its electronic alternatives. According to advocates, a bit tax could be collected by obliging operators to monitor digital transmissions.

Some European countries are showing disturbing interest in the idea. In

Belgium, for instance, a similar approach is apparently getting serious attention from the Telecommunications Minister who sees it as a means of redistributing wealth from the urban north of Belgium to the rural south. Under these proposals, the revenues from the bit tax would be used to provide social security rebates to labour intensive manufacturing companies.

Of course observers have been quick to warn about practical difficulties – different types of transmission would presumably have to be defined and different taxable value assigned. How is compression dealt with? How do you ensure that means are not found to 'smuggle' highly taxable applications into bit streams ostensibly transporting a low rated application.

Then there are privacy implications – under what conditions will monitoring take place and who gets access to the data?

At first sight, these initial proposals look unlikely to be adopted, but they are simply opening shots.

More proposals will surface as the on-line sector continues to grow and the potential 'take' looks more and more tempting to cash-stretched national administrations.

11

Cable Modems

The Internet's Worldwide Web is a fascinating development. In the short space of three years it has transformed the way all the players bidding for a power position in networked multimedia view their potential roles. Telcos and cable operators still have a pivotal position as providers of 'access' to the global network. But instead of driving and controlling the applications, as they assumed they would do four years ago by deploying high speed services, they are now running to catch up.

The global telecoms network across which the Web actually runs has become the weak link rather than the enabler. It is still weak at providing high-speed 'backbone' services at a reasonable cost. It is especially weak at connecting people to the Internet from their homes. Apart from ISDN, which remains a marginal access technology, the humble modem is still the only route onto the Internet for most users.

Telecoms operators must now compete to connect users.

We briefly looked at cable technology in Chapter 3. Now this chapter outlines some of the issues involved in 'migrating' the cable system towards a switched environment – where the users/viewers get to interact more and more directly with the content itself, as and when it becomes available, and as and when the user/viewer becomes culturally adjusted to using interactive service. Many of the UK's cable operators are gearing up to provide a first step towards the ultimate goal of a switched service environment using a technology called a cable modem.

To fibre or not to fibre?

Once the idea had been accepted that digital broadband interactivity was the destination of the network's commercial applications, the debate began to turn more and more around how much functionality should be built into the cable network, as it were, in advance. Should a bullish view be taken on capital investment so that the network was 'future proof', or should it be built (more economically) simply to deliver broadcast services to start with – with an upgrade remaining a possibility for the future?

The crux of this argument is about fibre and whether it should go right into the home or, if it shouldn't, how close to the home it should be taken.

High-fibre advocates believe fibre will ultimately be the key to any really 'serious' switched application (where each user has a unique channel or stream of data instead of sharing a broadcast) and the best approach should involve doing the job 'properly' from scratch, with high-speed fibres going all the way to the home or business premises.

High fibre advocates claim that taking fibre all the way will be cheaper in the long term, and will help to 'drive' broadband applications in the short.

Incrementalism

The opposing camp champions the idea of incrementalism. This argument says that most of the cost involved in building a cable network relates to the digging and ducting. Once the ducts are in the ground, more fibre can be pulled through as required. In any case, the technical scene moves so fast that it is not even certain that fibre will even be required for the final link.

Radio as a final drop for high bandwidth services is a distinct possibility; and powerful technologies have evolved which allow huge amounts of data to be pumped over copper. In the final analysis it may well end up being better to keep the final yards of copper 'drop' in place and upgrade the electronics when and where it is required.

Incrementalists point out that it is simplistic to think of there being a single technical solution for broadband in any case – the type of services required by small businesses and large businesses will be different; as will the types of services demanded by residential users.

Then there are simple physical and regional factors. High rise housing may demand a different distribution solution to low density, semi-rural dwellings. Differences in terrain may make a distribution solution impossible in one location, but economically viable in another.

Political differences

The high-fibre argument often has at its starting point the idea that high bandwidth services could and should be made available as a universal service in the interests of social equity. Incrementalism assumes that demand, at least in the initial stages, will be much more patchy. If the business is driven solely by commercial criteria, the network operator will be able to roll out high-speed pipes on a selective basis to those who are prepared to pay for them.

Incrementalists also tend to be less sure about the need for symmetry in the pipe (where data is designed to flow equally in both directions). They think it unlikely that substantial numbers of people will be in the business of sending vast amounts of data 'out' of the home. Apart from video-telephony, the only applications they feel it is possible to see clearly are inherently asymmetric, where nearly all the data comes 'in'.

High fibre advocates tend to be more certain that symmetry will be required because they expect that people will no longer be passive consumers of information and entertainment. For at least part of the time they envisage users distributing as well as consuming, because they also expect many people will be working over the network as well as being entertained by it.

Competitive pragmatism

A solidly commercial environment tends to err on the side of caution when it comes to future-proofing the basic technology. A certain amount of forward-thinking is certainly required, but push the assumptions too far out and investors tend to get nervous since, as we've seen, there are a lot of as-yet unquantifiable variables in all the calculations.

Unsurprisingly, the UK cable industry is incrementalist, and a broad consensus has been built around a pragmatic hybrid of fibre and coax copper cable as the basic infrastructure.

Fibre has the basic advantage of being able to pump data at high speed over long distances without picking up interference of any kind, but it is relatively more difficult to connect and requires more expensive equipment to terminate it at the home. As the technology currently stands, fibre is the most sensible medium for the distributing signals to the local 500 home nodes.

The coaxial copper medium has almost inverse characteristics – it performs best over short distances and is cheaper to terminate.

The immediate economics dictated that the best approach was to use both media where each was best suited. Fibre brings the TV signal to the local node, and coax makes the final connection. The network will then evolve as and when required.

Telephony

Pragmatism also ruled when it came to designing the cable telephone network. In 1990 the only alternative was to build a telephony overlay network. In other words, to build a completely separate telephone network in the ducts. Now, with cable the alternative telecoms infrastructure champion world-wide, equipment vendors are pushing completely integrated solutions, where broadcast signals and two-way telephone share the same physical media.

It is possible that some of the networks laid during the last half of this decade will incorporate an integrated approach, although the jury is still out over its real cost and reliability

In the end, the high fibre argument is only partly a technology squabble. The high-fibre view often has a national interest dimension, and the most ambitious 'fibre-the-nation' plans usually involve substantial government investment. Italy and Japan are both (at least at the time of writing) going down this road. In these circumstances, the technical arguments in favour of fibre to the home or to the curb are often overlaid with overtly political motives – like creating jobs or galvanising a sense of national renewal (in the old white heat of technology mould).

The broadband multi-service network is also the ultimate destination of the telco network too. So it should come as no surprise that a high fibre approach is often advocated most vociferously by incumbent telcos who see clearly that a 'strategic' dimension to the infrastructure (beyond that which would be supported by private capital) would entrench their positions as national infrastructure providers, especially as such plans almost inevitably involve massive cash injections from the state.

The overt logic behind this argument is to 'rationalise' infrastructure investment through a unitary network. This would, for all intents and purposes, retard the development of broadband infrastructure competition, because even were it allowed, given the level of government support for the incumbent telcos they would still have a huge advantage.

Needless to say, grand plans of this sort are not popular in the cable sector.

While BT is currently excluded from delivering broadcast entertainment, it will eventually be unleashed on the basis that sustainable competition in the telecoms market has been attained.

BT would like to have the option sooner. Its current policy position seems to involve dangling the political carrot of a major broadband network investment in front of Labour politicians, while using a populist stick – bemoaning unfair foreign competition and harsh restrictions – to beat Tory ones.

It maintains that without the ability to deliver broadcast entertainment it is unable to make the sums add up to justify the investment. Therefore, unlike other countries, UK effectively lacks an IT 'champion' to propel it into the information society.

Uncomfortably for the cable industry, this line seems to have struck a chord with the Labour party which unwisely announced a tentative 'deal' with BT should it attain office. It would release BT from its broadcasting ban and BT would offer free service connection for schools and hospitals.

Many in the cable industry suggest that BT's primary motive, both in agitating for the ability to broadcast and in trialing interactive video-on-demand (VOD) services over its telephone network, is to weaken long-term investor confidence in the cable industry.

To be fair to BT, while this might have come as a welcome secondary effect, BT was hardly alone amongst the world's telcos in trialing interactive services. With the company still in possession of a large research and development facility to guide it to future opportunities, this relatively minor experimental effort had every justification.

In fact, given the level of interest in interactive service developments, BT's own shareholders should have been rightly worried if it have decided not to explore the options.

But BT's position on broadcast is slightly ironic. Given that it took the cable industry just four years in telecoms to see its broadcast revenues outstripped by its telecoms ones, BT's ability to come late to broadcast entertainment hardly looks likely to make or break the economics of broadband. However, BT's Chairman, Sir Iain Vallance, maintains that the ability to broadcast is an essential part of the whole puzzle, in that it would allow BT to compete successfully for content and would provide an important part of a broadband 'package' to the consumer.

Cable modems

One of these increments may be represented by the cable modem. This is a gadget which connects the coaxial cable which the cable operators use to feed the television signals into subscribers' homes to a PC, network computer, or in the future, a Worldwide Web-enabled TV.

The cable modem tunes in to one of the frequency bands normally used to deliver a television channel and is able to decode a data stream of between 10 and 30 million bits a second being carried on the channel.

In fact the term 'cable modem' is not really accurate as the box is more technically similar to what is called a bridge/router in the local area network sector. Its function is not just to decode the data, but to sift it for those packets addressed specifically to it and forward these to a program running on a PC or terminal of some kind.

The cable modem is a perfect example of the sort of small increment that the cable network is able to support. It doesn't provide an over-arching switched data service, but it does seem to provide the perfect way of connecting subscribers to the Worldwide Web.

Technical background

As the cable network was engineered to support a one-way, broadcast system, getting data down it using one of the TV channels is relatively easy. Getting data to flow 'back' the other way is more difficult to engineer without also introducing electrical interference which may 'leak' into the network from users' premises.

As a result cable modem design favours an asymmetrical approach – where high bandwidth is delivered downstream and controlled by a narrow upstream channel, operating at low frequencies, across which users can send commands or return relatively small (in the scheme of things) email messages.

But this, it now transpires, is just the sort of underlying capability required to support Worldwide Web access, where users are generating tiny datagrams containing a handful of characters to pull down increasingly large pages, often containing tens of thousands of characters of graphics material.

Cable modem services may also be capable of delivering video-on-demand and other, more synchronous services in the future.

The cable companies, like the world's large telcos, BT included, have now come around to the idea that Internet access service is a game they should play. The cable companies, several of whom have launched their own conventional Internet access services, see the cable modem as a possible way to 'differentiate' themselves from a growing host of competitors with similar dial-up options. With the cable modem in their armoury they hope they can develop a service which can appeal immediately to the 'heavy user' community.

According to Telewest, these 'heavies' comprise a significant proportion of internet users and they already spend around £40 or £50 per month on standard access and telephone charges. The operators calculate that a cable modem service might be priced to come in either at, or slightly above this and would offer significant performance and ease-of-use advantages.

Asymmetrics and the cable modem

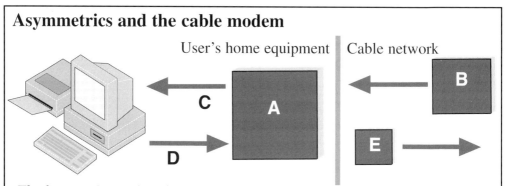

The home PC or other device uses an Ethernet LAN-style interface (A) to 'listen' to the data being carried 'downstream' over the shared channel (B). As with a LAN-attached workstation the interface identifies packets of data which are addressed to it and forwards them to the computer and its program (C). Upstream data (D) is carried by the low speed cable return path (E) operating at another frequency range.

Sharing the pipe

Unlike the existing 'dial-up' methods of gaining residential or small business Internet access through the standard telephone line or via ISDN, with a cable modem the user is effectively using a very small share of a very large piece of bandwidth, just as the same user might do when connected to a local area network (LAN) at the office. This shared bandwidth approach brings with it a whole host of advantages for both user and service provider.

For the user there is no slow and tedious 'dial-up' procedure. Instead the cable modem could provide what the telecoms industry calls a 'permanent virtual circuit' between the Internet access node and the user's computer applications (at least while the computer is fired up and running the relevant software). So access software (browser, email and perhaps others as they develop) can be used on demand and incoming email can be flagged up on the screen. In short, the cable modem seems to offer all the advantages of being connected to a corporate LAN with an Internet gateway.

The cable companies also expect that installation and set-up can be made much more coherent and trouble-free than is currently the case when users subscribe to an access service via dial-up modem or ISDN.

The shared high bandwidth also brings a huge increase in speed for users' applications. Instead of a bottleneck forming at the 'access network' (conventionally, the local telephone line) the speed at which Web pages can download will be limited by the speed of the Internet itself.

If cable modem service delivers on its initial promise the cable operators may be in a position to price service for a mass market, rather than just concentrate on the Internet high spenders.

Telco competition

Sooner or later competition in the high-speed Internet access market will come from BT, which will be able to deploy a similar sort of technology, but engineered to make the best of the telecoms network. These telco technologies, known as HSDL (High Speed Digital Loop) and ADSL (Asymmetric Digital Subscriber Loop) use roughly the same sort of advanced modulation approach deployed by the cable modem, to supercharge the simple twisted copper pairs which usually carry the telephone circuit from the local exchange to the customer. This technology is well proven. HSDL is already used in the telecoms network to support multiple voice channels down a single copper pair. ADSL is currently being used by BT in its video-on-demand trial in Colchester.

Like the cable modem ADSL is also benefiting from cheaper, more integrated electronics and vendors are now beginning to announce products that telcos can deploy to provide high speed services.

But the cable modem has a few advantages. From the operator's perspective its ability to share bandwidth between several users allows high speed services to be deployed relatively economically. Once the cable operator has engineered the required equipment at the 'head end' (the hub of the cable network where the cable is fed its programming) there is only an incremental cost associated with rolling out service to each existing cable user who takes it – no upgrade of the existing physical network is required. Turning an ordinary telephone line into a high speed digital line often requires some testing and line conditioning to ensure it functions smoothly – all this adds to the cost of deployment.

A cable modem service is also flexible. In rather the same way that capacity is increased on a LAN by deploying filtering bridges, the cable network can be 'segmented' to have different streams of data shared over smaller geographical areas as usage grows. Extra channels can also be turned over to cable modem traffic to meet demand, or to provide different classes of service.

Of course, a cable modem service would quickly move the Internet bottleneck upstream into the Internet itself well before the user is enjoying anything like true 10Mbit/s downstream access. The cable companies claim that they have plans to offer even more value by caching popular, local, or data-heavy material at their 'head ends' so that it can be delivered at full speed to make the most of the service.

Initial cable modem services are likely to come with the box as part of the deal. But some pundits are predicting that eventual standardisation, and therefore commodity pricing, will see cable modems being sold direct to consumers. At this point cable operators could offer a cheaper service where the user provides his own modem, or a more expensive service where he doesn't.

Industry enthusiasm for the cable modem concept in the US is certainly oiling the cogs. There cable modem service is currently expected to be a huge market, given the larger cable TV penetration and relatively higher Internet enthusiasm in North America. Many big name communications equipment vendors have, or are planning, a cable modem offering and the first stages of mass production are now beginning.

Good timing

The cable modem appears to be the right thing at the right time. Internet access suits an asymmetrical service – much capacity required downstream, only a narrow channel required back.

The current generation of cable modems are also asymmetric, partly because Internet access is the obvious application and they're being designed that way, and partly because the upstream channel presents technical problems.

However symmetrical (or at least more symmetrical) products can also be deployed where they need to support applications like home-working and remote office LAN connection where the data flows equally in both directions.

ISDN alternative

As an alternative to ISDN access, therefore, the cable modem may have several advantages.

The first is obviously speed. ISDN channels deliver fixed amounts of bandwidth in multiples of 64kbit/s, cable modems will be able to offer throughput in the hundreds of kilobits range and possibly higher. From the operator's standpoint, modem service may also be economical to deploy – in comparison to basic rate ISDN – as it uses the existing CATV transmission network. Head end support will carry a cost, of course, but roll-out costs will be incremental. This is important in the early stages of deployment when penetration is thin.

Cable modem enthusiasts also claim that they will be easier to install and maintain than clumsy basic rate ISDN services, due to their network management facilities.

Technology issues

There are many issues to be hammered out. To begin with, the technology is still at its proprietary stage. While standards-setting is under way, past experience suggests that the process always takes longer than originally thought.

There is currently much confidence-building going on in the cable modem field with some researchers predicting commodity pricing by 1998. This may turn out to be accurate, but there is always a tendency for the equipment supply side of the industry to 'talk up' new technologies, aided and abetted by market reports and favourable press coverage. If confidence is not generated, new technologies tend to be caught in a 'Catch 22', where a small market equals high equipment pricing, equals a permanently small market. A confident momentum is required to turn this into a virtuous cycle, where building-block vendors (silicon manufacturers and the like) feel confident enough to plan for a mass market and are prepared to price their product to build market share in the early stages.

Trials

So far, the cable industry has been relatively low-key about the possibilities, but several of the larger Multiple Systems Operators (MSOs) have been conducting small trials of the available technology for some time, and now the largest UK MSO, Telewest, is planning to offer high speed Internet access to some of its customers this year. Telewest has already been trialing applications in schools where participants are connected to a central CD-ROM server to provide interactive digital content. Other services or trials are also expected

Major movement

In the US, by late 1996, commercial deployment of cable data services, using cable modems, was beginning in earnest. US MSO, Continental Cablevision, had agreed to purchase 50,000 LANcity modems to support a fully commercial, two-way high speed Internet access service available to customers along Boston's Route 128 corridor.

As the East Coast's answer to California's Silicon Valley, Route 128 offers the ideal Internet access demographics and Continental Cablevision has upgraded its systems to two way hybrid fibre/coax to support the service. But the MSO is also targeting a more representative US territory in Jacksonville, Florida.

Here it is also rolling out 10,000 General Instrument cable/telephone hybrid units which use a standard dial-up modem for the back channel. A high speed downstream option will be provided as an upgrade to the company's recently launched standard dial-up Internet access service.

Marketing strategies

US cable operator strategies for data services are markedly different from those likely to be pursued in the UK.

US cable penetration rates of up to 60% provide a different starting point, but the US environment also exhibits some disadvantages when compared to the UK. With a lot of narrowband cable still in the ground, US operators will be faced with the problem of providing hybrid solutions involving telephone back channels.

But perhaps most crucially, US cable Internet access must be marketed as a premium service. Continental Cablevision plans to offer unlimited high speed Internet Access at between US$30 and US$45 per month. But of course, in the North American market local call costs are comparatively low or even free, making it that much more difficult for cable data service to match dial-up on cost.

In the UK, while local call costs are falling fast, they're still high enough to see a significant number of 'heavy' dial-up Internet users spending as much as £30 or even £50 per month on telephone charges and ISP subscription.

With the high level of US cable data activity, commodity pricing is kicking through to cable modem deployment costs and this is already affecting the strategic thinking of those planning the services here.

While the UK cable industry is still at the early planning stages, the option to go for high penetration with US-level pricing (between £20 and £30 per month, say) looks more and more attractive.

But even at higher prices, cable data services would still be an attractive proposition for a 'heavy user' because it could offer 'quality' service for the same real price as dial-up access.

12
The Preferred Pipe

Investment in the UK cable business is really a long-term bet on the ultimate worth of a broadband network – that however the entertainment, information and communications industries develop, a broadband, fully interactive network is most likely to be able to deliver and price a range of services which can compete, in combination, with individual services on offer through alternative delivery methods.

From this perspective cable is a 'long haul' investment. With very heavy up-front capital investment, the task will be to build a diversity of profitable services which can then amortise those deployment costs. These may be delivered as a package if market advantage can be found in doing so. But at any rate, as the range of solutions required by an increasingly diverse market grows, cable operators are banking on their having a marginal cost advantage as each one is introduced.

If all the hopes held for cable modems are borne out for instance, this data service delivery technique will show its real competitive teeth by having an incremental cost of deployment in what will prove to be a very cost conscious market (see Chapter 11).

The ultimate aim will be to become the preferred pipe into the home and business – a more powerful proposition as the range of home delivered services continues to grow.

In the meantime, competitive delivery methods may be able to engineer short term advantages by tailoring a network to deliver a particular service. In this context, cable's residential and small business telephony base will face competition from Ionica and other radio operators.

Ionica's approach is doubly interesting because in the UK it represents an outing for the opposite set of long-term assumptions.

Ionica's advantage is that it has a relatively incremental cost of deployment per customer. It doesn't have to 'pass' three or four customers for every one who takes service and it may even spend less connecting every customer who does.

But of course, this is exactly the cable assumption when it comes to the next generation of services. The network is in the ground and adding another revenue stream to it should represent a marginal cost.

If you have three or even four families of services running across the same infrastructure, then you have a host of advantages into the future.

Fragmentation

The so-called 'fragmentation' of the cable industry can be a strength in that it allows divergent approaches to be tested as a natural by-product. Both formal and informal networks exist at all levels within the cable business community, so mistakes and successes can be avoided or emulated.

Writer George Orwell maintained that "everyone's life, when viewed from the inside, is simply a series of failures". This may say more about Orwell, who was a particularly driven individual, than it says about the human condition. But if an anthropomorphous modern enterprise were to be sketched its psychological profile would resonate to the attitude implied here.

When viewed from the inside, every enterprise appears more chaotic and inefficient than it might appear from a distance. Certainly, the way the cable industry, in its fragmented way, approaches its market can seem to make the management task more difficult – especially when competing with a monolith.

On a day-to-day basis the advantages of small-size, focus and manoeuverability may understandably be subjectively experienced as their corollaries – tight financial controls, lack of breadth and lack of direction.

But cable's structural diversity has allowed it (forced it) to create some interesting structures. These horizontal companies – formed to undertake specific functions across 'vertical' franchises might also be one direction in which the fragmenting telecoms industry as a whole is headed.

Cooperation

The logic of sharing research and development effort, for instance, saw Bell Cablemedia, Nynex and Telewest team up to establish a media laboratory – launching the first phase of the multimedia services trial agreed between the three companies early in 1996. Its objective was to establish compatible standards and services to allow the cable industry to provide a seamless, nationwide set of interactive multimedia services to customers and content providers. But this initiative was quickly overtaken by events. By mid-1996 it had become clear that the Internet and cable modem technology should be the focus for multimedia development.

Sharing technology

In 1996, the industry had also formed a Cable Modem Group through the offices of the Cable Communications Association (CCA), the industry's joint body. The group will be comprised of representatives from interested cable operators and its aim will be to look at sharing information and, where possible, formulating some joint policy aims. These are likely to include technology and service standardisation.

Cable's cable

The London Interconnect Group is also an example of the kind of 'horizontal' organisation the cable companies might be expected to develop to meet specific needs.

Founded in 1994 by a consortium of six cable operators with a presence in the London metropolitan area: Bell Cablemedia; Cable London; The Cable Corporation; Nynex, Telewest and Videotron, the LIG interconnects the operators across a high-speed SDH network. This allows them to exchange switched calls passing between their networks but, more importantly, it has created a way to provide business services on a London-wide basis without using an intermediary telecoms carrier to establish the connections.

Cable companies have also formed a Northern Interconnect group to perform similar functions across the important conurbation comprising Liverpool, Manchester, Leeds and adjoining towns and cities. One possible scenario involves the interconnect organisations gradually extending their range and then interconnecting with each other to provide one route for national cable traffic.

To win the important medium to large business telecoms customers, the cable companies needed to match BT or Mercury's ability to provide end-to-end connections for leased circuits (or virtual leased circuits). This was not simply a technical requirement.

What was also required was the ability to present a unified commercial face to the target customers – a single point of contact through which the services could be arranged and, importantly, through which they could be charged out on a coherent basis.

The LIG established a commercial arm called ICn (Integrated Communications Network). ICn is charged with the job of acting as the sales and technical 'glue' between the London cable operators so that the metropolitan capability can be stitched into place.

As well as providing the technical facilities the LIG establishes some procedures through which joint projects can be steered. Through the LIG the potentially knotty problem of how best to approach potential customers who straddle two or more MSO territories (local authorities, for instance) can be decided.

Sales forum

It also serves as a platform for other forms of co-operation. It ultimately makes sense to market both business and residential services on a joint basis if and when it is possible to arrive at a common package – at least at the entry-level – involving both TV and telecoms residential services.

One obvious way to drive penetration in the residential market may be to establish free calling (in the evenings, say) between cable customers. The division of the metropolitan areas between operators obviously blunts this approach, but through the interconnect, operators would like to be able to offer free calling offers on calls to all other London cable subscribers – exponentially increasing the attractiveness of cable service and introducing 'word-of-mouth' marketing, where friends and family encourage each other to take cable service to enjoy free calling.

The Interconnect organisations are also playing a similar role on the business side. The logic of Centrex, for instance, can be leveraged by cable company co-operation across a metropolitan area.

The high-speed network will increasingly distribute TV as well. It currently carries 20 digital TV channels encoded at 190Mbit/s The ICn network distributes the London-based 24 hour rolling news channel - Channel One.

The LIG facilities may well be able to be migrated to rationalise the head-end functions now currently undertaken by each operator. Instead of each local distribution networking intercepting satellite channels at each head-end from a dish and pumping it into the local cable system, the economics of transmission may move to a point where both cable-only and satellite channels could be pumped across the high speed interconnect backbone.

This may look more economically feasible as broadcast TV moves to digital format.

Buying into the backbone

In the same vein Cabletel acquired NTL, the privatised transmission network arm of the UK's Independent Broadcasting Authority. NTL's core business is to distribute ITV, Channel Four and other broadcast media to transmission sites around the UK. Since its privatisation NTL has expanded its activities into the telecoms market. It has a long-standing agreement with the cellular operator, Vodafone, to provide backbone services and last year it signed up Orange. The synergy between a microwave network operator and a cellular operator are interesting – when the Orange deal was signed, the companies were already sharing towers for transmitter equipment.

The companies say that together they can provide a unique 'end-to-end' broadband network by marrying Cabletel's fibre-based local networks with NTL's SDH-based broadband national backbone. Bringing NTL into the cable 'fold' in the UK will provide some important new business for NTL – carrying cable telephony traffic between franchises or to BT interconnection points.

For Cabletel, the acquisition promises to open up a host of opportunities. as the network can be an important component in interactive service delivery.

Targeting broadcast

The new interactive business may not rely totally on the Internet. It should be possible to package interactive service with Cable TV content so that specialist content providers can build more tiers of value into their offerings. The cable companies would be in a position to sell access to interactive versions of the same content as an add-on.

Weather services represent an interesting pointer to the way such a process could develop.

Two programmers are now competing over the 'weather market' in the UK. Both companies are planning rolling weather forecasts with the ability to focus on specific regions – the weather can apparently be specified down to around a 30 mile radius. With its ability to deliver geographically-based audience segments, cable (as opposed to satellite) makes the ideal delivery mechanism for this level of geographic segmentation.

The Weather Network claims that it has been developing its format for a year and has an advanced all-digital facility to provide the feeds. It will be bouncing a multiplexed digital service from satellite from day one. Specific regional services can therefore be demultiplexed, by a box provided, at participating cable operator head ends so that micro-forecast segments can be delivered directly to subscribers.

But the weather channel concept is also different in that it comes close to being a 'video-on-demand' service which happens to be delivered as a series of broadcast channels.

Unless you're very sad, a weather channel is not something you sit down and watch like rolling news. The idea is that you 'dip' into it when you want to know what the weather is going to be like and that you're never more than a few minutes away from a standard forecast.

Given the way these weather services are designed the obvious next step is to provide Worldwide Web content to go alongside it and offer even more specific or detailed levels of weather information. Armed with a cable modem service, cable companies would be in a position to provide another level of value for the same information need.

Data centrex

Having end-to-end control over a broadband access network also presents a host of opportunities in the business market over the next 10 years or so. If IP, the Internet's protocol, becomes the standard for handling the Network Layer – the set of rules under which a network of computers exchanges streams of data – the resulting homogeneity of users' data networking requirements should put cable companies in a position to offer central office switching for local area networks (LANs).

This scenario would involve the crucial hubbing, routing and perhaps even data storage functions currently carried out by boxes on the users' premises being taken over by the broadband telecoms operator.

As things currently stand, IP does look like staying the course. While competition between Microsoft and Netscape may fracture the Worldwide Web itself into proprietary camps, the underlying IP running the Internet will remain in place whoever wins. And IP is going to remain the way most corporate computer networks communicate and arrange themselves. So whatever specific technologies telcos find most suitable in the public network, be it ATM, SMDS or even SDH pipes, it is going to have to have IP mapped onto it. That, for the foreseeable future, is what many customers are going to want, especially if they continue to build so-called 'intranets' which allow Internet technologies to be used across a private network to support information distribution and collaborative working. Judging by the amount of support the intranet concept is currently getting, the communications equipment and computer vendors clearly think they will.

If voice Centrex scores hits in the corporate market because it allows users to avoid owning a switchboard, how much more attractive a data version of the same commercial proposition might appear to a beleaguered data network manager. At present he feels he spends too much of his time ironing out minor technical glitches amongst his shelves of flickering hubs and routers, instead of getting on with managing the networked applications.

A data version of Centrex could see telcos and cable companies connect the desktop to their switching centres. With IP as the network's *lingua franca* they would have a unified basis on which to develop switching technologies and systems. Using these they could coherently manage dozens of virtual LANs (especially if they supported users in different buildings), adding the resilience, security, and reliability that the corporate network manager currently finds such a struggle to attain.

Prospects

Much is currently expected of the global cable industry. With telecoms liberalisation under way in the US and full infrastructure competition looming in much of Europe, cable has had a new role thrust upon it – as the missing local link in the evolving blueprint for full-network competition. As a simple entertainment delivery business in Europe and North America the cable industry has spent several decades developing slowly and undramatically as a fragmented set of utilities – tightly regulated, capital-intensive, long term and irrevocably 'local' in its scope and ambition.

But while an increasingly digital and flexible technology might seem to give the global cable industry the technology to compete with the world's incumbent telcos, it actually leaves the starting blocks with one fundamental disadvantage – its fragmentation.

Current UK experience suggests that when the telco fights back, as BT is now doing against the UK's telephony operating cable companies, the incumbent's ability to market itself nationally with an increasingly complex package of tariffs and features gives it a major advantage.

In the first phase of local loop competition in the UK, a strong hand appeared to be held by the new cable operators and they quickly built market share on a simple price proposition.

But as margins diminished, more as a result of price regulation than of real competition, both sides became anxious to move the competitive battle away from price and on to more sustainable ground.

BT finds its range

To this end BT has finally found its most effective advertising message – that its customers think their call costs are far higher than they actually are.

Telling them that they aren't only makes them more likely to make what they previously thought were expensive calls, but moves the cost issue down the scale of concern. It is early days yet, but if BT can reinforce the message that its prices are now much cheaper than they were, it must blunt the force of the competitive price propositions being aimed at its customers.

The cable industry, too, is mounting a joint national marketing campaign aimed at residential subscribers – the first phase of which reinforces the cost advantage which the industry as a whole must obviously maintain. But while the cable industry is able to develop a series of general messages, it is unable to focus on specific packages nationally because each Multiple System Operator (MSO) peddles a unique set of services and prices. Although the industry is consolidating into a handful of major MSOs, it is unlikely to become a single entity any time soon, if at all.

This is one minor skirmish in a long and complex battle, but it does illustrate how the scene changes as the marketing element assumes greater importance. What used to appear like major cable advantages – the companies' local focus, ability to act quickly, their second revenue stream from cable television, and their fibre rich network, no longer look quite as attractive in isolation.

The real fight will be around 'packaging' services for different categories of customer and getting the message across coherently. In this department, so far at least, BT, as a single national entity has a substantial advantage because it is inherently more able to deploy national marketing. Consolidation amongst cable operators will therefore continue and various kinds of 'horizontal' cooperation, involving national marketing or joint approaches to specialised services, will intensify.

Infrastructure competition wins

There was a time when telecoms policy-makers thought the competition god could be sated by the establishment of competing service and platform providers. Underneath, a single national network provider, dispassionately wholesaling capacity to the tiers above, might remain a monopoly .

But it is now almost universally accepted that local infrastructure competition must underpin the competitive edifice. Rival entities at every level are now accepted as a pre-requisite to the establishment of a virtuous cycle of innovation, revenue growth, and price competition. With infrastructure out of this loop the competitive engine would be running on only two cylinders. So it is vital that alternative access network operators are able to build substantial market share through the long transition to sustainable competition at all the levels.

But the UK experience is showing that balancing the competitive environment is no easy matter. Worldwide, regulators are now being urged to 'withdraw' from detailed regulation to a position of 'competitive oversight'. The language used implies that such an approach will place them, or at least the regulatory function, in a more serene, almost paternalistic role.

Competition, it is implied, will solve most of the taxing problems and the regulator will simply ensure that it is all fair and above board – a touch on the brake here, a flick of the indicators there, and a few slaps on the wrist all round. But umpiring that competition is likely to become an even more complex business.

The UK experience has shown just how important telecoms interconnect arrangements are if fair competition is to be sustained. Competition and market forces are not enough – a powerful regulatory hand is required. As the UK battle over number portability has shown, interconnection becomes more difficult to arrange as the services being interconnected become more complex. As competition builds, and time-to-market matters even more, the ability of the telecoms incumbent to use the regulatory process to simply slow down the opposition becomes a competitive game in its own right.

The preferred pipe

The initial phase of telecoms competition, where the cable companies have been a major force, has seen UK telecoms prices driven down and service levels driven up as BT has responded.

The next phase will see the introduction of advanced business services, like Centrex, and interactive residential offerings. Where the conventional telecoms market saw cable competing against a monopoly incumbent, the next phase will often involve the cable companies getting to market first.

The next phase will also see a more even playing field. With over half the cable industry now consolidated as Cable and Wireless Communications and more consolidation expected, the industry now has the size and scope to compete effectively with BT.

Centrex for small- to medium-sized businesses, where many cable companies are currently concentrating their efforts, is almost virgin territory in the UK. Broadband residential services, enabled by the cable modem, will be introduced in 1997, well ahead of alternatives from BT.

With multi-channel TV, telephone and interactive services, the UK's cable industry will have the ability to position itself as the 'preferred pipe' for their business and residential customers.

Glossary

Analogue
An analogue (US, analog) signal is a concrete representation of a 'real-world' phenomenon – for instance, in analogue voice communications a fluctuating voltage signal generated by a microphone is an analogue of the sound waves which have caused the diaphragm in the microphone to vibrate. Hence these signals form an analogue circuit. A digital signal, on the other hand, would deliver an abstract definition of the same sound waves as a series of numerical values in binary.

Asymmetric
A communications system designed to allow more data to flow in one direction than the other (usually downstream to the user).

Asynchronous communications
Where data is transmitted so that each character or byte (of 7 or 8 bits) is sent separately instead of forming part of a synchronised stream. The receiving device samples the signal, say every 3-thousandths of a second, following the arrival of a 'start' bit. After 8 3-thousandths of a second samples it has decoded the character. It then waits for the arrival of the next 'start' bit to begin its timing again. Asynchronous transmission is both cheap and reliable and is used by dial-up modems. Asynchronous Transfer Mode is the same principle applied to self-contained packets of data which travel across the network independently.

Bandwidth
The range of frequencies a transmission medium can carry: this is

essentially an 'analogue' term (see Analogue) and relates to the size of the band of frequencies available for carrying information: the greater the bandwidth, the greater the information-carrying capacity of a channel.

Bit
One unit within the binary numbering system, either a '0' or a '1' – is a hybrid of BInary digiT.

Bits per second
The rate at which bits are transmitted across a communications link; written bit/s. 1024 bit/s is 1kbit/s (kilobits), and 1 million bit/s is 1 Mbit/s (megabits).

Broadband
A relative term relating back to 'bandwidth'. Broadband has now come to simply mean fast. So with cable TV it describes the ability of the cable system to carry 30 or more TV channels. In telecoms, broadband now simply refers to digital technologies or services running at greater than 2Mbit/s.

Broadcast
The simultaneous transmission of signals to a set of destinations or to all destinations. Broadcast means one-to-many, as in broadcast TV.

Byte
Eight bits forming a unit of data. Usually each byte is one character.

Circuit switching
Telecoms originally involved setting up a physical electrical circuit between sender and receiver. As digitalisation was introduced, a circuit came to mean a channel set up across the network guaranteeing the end-to-end transmission of a fixed amount of data for the duration of a call, usually 64kbit/s. While this no longer involves 'circuits' in the electrical sense, the service is the same from the user's point of view.

CPE (Customer Premises Equipment)
Telecoms equipment, including PBXs and wiring, located on the user's premises.

DASS II (Digital Access Signalling System)
BT's signalling method which allows PBXs to control a primary rate ISDN interconnection to the public network. DASS allows the pbx to tell the network where to send a call, when to disconnect it and so on.

DDI (Direct Dial In)
Where telephone extension users in an organisation are assigned a full telephone number, allowing callers to dial directly to the extension without having to go through the organisation's switchboard operator.

Digital signal
A signal with only two values, normally 0 and 1. In contrast to an analogue signal whose values vary (see analogue).

Head end
Associated with CATV technology, the head end is the control centre for a cable system where signals are received, processed and sent for distribution down the cable system.

HFC (Hybrid Fibre/Coaxial)
A CATV topology comprising fibre optic cable to the street cabinet supporting up to 500 homes, followed by coaxial cable to the premises. HFC equipment can deliver voice as well as broadband signals over the single cable. HFC is the preferred architecture for many CATV operators aiming to deliver interactive services.

Microwave
Wireless transmission at very high frequency to deliver telecoms services, including TV distribution, between two points. Microwave transmission is dependent on line-of-sight.

Modem
Amalgam of the words MODulator and DEModulator. Modulates an outgoing binary bit stream onto an analogue waveform, and demodulate the incoming equivalent into binary.

Multiplexer

A device that can amalgamate several communications channels onto a single circuit.

MVDS (Microwave Video Distribution Service)

A TV distribution service using microwave transmission to the home.

PBX or PABX (Private Automated Branch eXchange)

Private (usually corporate) telephone exchange or switchboard linked to the PSTN.

Packet

A collection of bits, including the address, data and control, that are switched and transmitted together. The terms frame and packet are often used synonymously.

PSTN (Public Switched Telephone Network)

A public telecommunications network using the conventional 'dial-up' method of connecting one subscriber to another.

Virtual circuit

A network service that behaves to the attached systems as if it were a physical circuit – the signal or data sent in at one end, comes out at the other as if the two were connected.

Index